W9-DDK-085

SKILLS MASTERY

This Book Includes:

- Practice questions to help students master topics assessed on the the PARCC and Smarter Balanced Tests.

  ▸ Ratios & Proportional Relationships
  ▸ The Number System
  ▸ Expressions & Equations
  ▸ Geometry
  ▸ Statistics & Probability

- Detailed Answer explanations for every question
- Strategies for building speed and accuracy
- Content aligned with the Common Core State Standards

Plus access to Online Workbooks which include:

- Hundreds of practice questions
- Self-paced learning and personalized score reports
- Instant feedback after completion of the workbook

Complement Classroom Learning All Year

Using the Lumos Study Program, parents and teachers can reinforce the classroom learning experience for children. It creates a collaborative learning platform for students, teachers and parents.

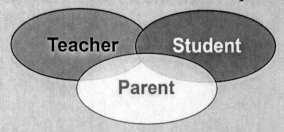

Used in Schools and Libraries
To Improve Student Achievement

Lumos Learning

Common Core Practice - Grade 6 Math: Workbooks to Prepare for the PARCC or Smarter Balanced Test

Contributing Author	-	**Renee Bade**
Contributing Author	-	**Kimberly G.**
Curriculum Director	-	**Marisa Adams**
Executive Producer	-	**Mukunda Krishnaswamy**
Designer	-	**Mirona Jova**
Database Administrator	-	**R. Raghavendra Rao**

ISBN-10: 1940484464

ISBN-13: 978-1-940484-46-4

Printed in the United States of America

For permissions and additional information contact us

Lumos Information Services, LLC
PO Box 1575, Piscataway, NJ 08855-1575
http://www.LumosLearning.com

Email: support@lumoslearning.com
Tel: (732) 384-0146
Fax: (866) 283-6471

Lumos Learning

Table of Contents

Introduction

The Common Core State Standards Initiative (CCSS) was created from the need to have more robust and rigorous guidelines which could be standardized from state to state. These guidelines create a learning environment where students will be able to graduate high school with all skills necessary to be active and successful members of society, whether they take a role in the workforce or in post-secondary education.

Once the CCSS were fully developed and implemented, it became necessary to devise a way to ensure they were assessed appropriately. To this end, states adopting the CCSS have joined one of two consortia, either PARCC or Smarter Balanced.

Why Practice by Standard?

Each standard, and substandard, in the CCSS has its own specific content. Taking the time to study and practice each one individually can help students more adequately understand the CCSS for their particular grade level. Additionally, students have individual strengths and weaknesses. Being able to practice content by standard allows them the ability to more deeply understand each standard and be able to work to strengthen academic weaknesses.

How Can the Lumos Study Program Prepare Students for Standardized Tests?

Since the fall of 2014, student mastery of Common Core State Standards have been assessed using standardized testing methods. At Lumos Learning, we believe that yearlong learning and adequate practice before the actual test are the keys to success on these standardized tests. We have designed the Lumos study program to help students get plenty of realistic practice before the test and to promote yearlong collaborative learning.

This is a Lumos tedBook™. It connects you to Online Workbooks and additional resources using a number of devices including android phones, iPhones, tablets and personal computers. Each Online Workbook will have some of the same questions seen in this printed book, along with additional questions. The Lumos StepUp® Online Workbooks are designed to promote yearlong learning. It is a simple program students can securely access using a computer or device with internet access. It consists of hundreds of grade appropriate questions, aligned to the new Common Core State Standards. Students will get instant feedback and can review their answers anytime. Each student's answers and progress can be reviewed by parents and educators to reinforce the learning experience.

How to use this book effectively

The Lumos Program is a flexible learning tool. It can be adapted to suit a student's skill level and the time available to practice before standardized tests. Here are some tips to help you use this book and the online resources effectively:

Students

- The standards in each book can be practiced in the order designed, or in the order of your own choosing.
- Complete all problems in each workbook.
- Use the online workbooks to further practice your areas of difficulty and complement classroom learning.
- Download the Lumos StepUp® app using the instructions provided to have anywhere access to online resources.
- Practice full length tests as you get closer to the test date.
- Complete the test in a quiet place, following the test guidelines. Practice tests provide you an opportunity to improve your test taking skills and to review topics included in the CCSS related standardized test.

Parents

- Familiarize yourself with your state's consortium and testing expectations.
- Get useful information about your school by downloading the Lumos SchoolUp™ app. Please follow directions provided in "How to download Lumos SchoolUp™ App" section of this chapter.
- Help your child use Lumos StepUp® Online Workbooks by following the instructions in "How to access the Lumos Online Workbooks" section of this chapter.
- Help your student download the Lumos StepUp® app using the instructions provided in "How to download the Lumos StepUp® Mobile App" section of this chapter.
- Review your child's performance in the "Lumos Online Workbooks" periodically. You can do this by simply asking your child to log into the system online and selecting the subject area you wish to review.
- Review your child's work in each workbook.

Teachers

- You can use the Lumos online programs along with this book to complement and extend your classroom instruction.

- Get a Free Teacher account by using the respective states specific links and QR codes below:

PARCC States	SBAC States
LumosLearning.com/a/stepupbasic	LumosLearning.com/a/sbacbasic

This Lumos StepUp® Basic account will help you:

- Create up to 30 student accounts.
- Review the online work of your students.
- Easily access CCSS.
- Create and share information about your classroom or school events.

NOTE: There is a limit of one grade and subject per teacher for the free account.

- Download the Lumos SchoolUp™ mobile app using the instructions provided in "How can I Download the App?" section of this chapter.

How to Access the Lumos Online Workbooks

First Time Access:

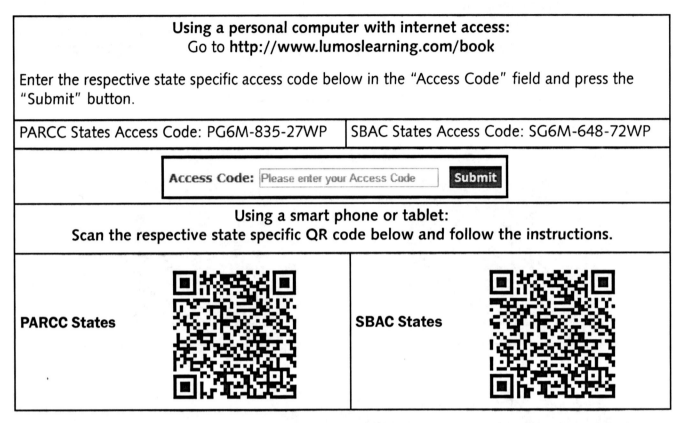

Using a personal computer with internet access:
Go to **http://www.lumoslearning.com/book**

Enter the respective state specific access code below in the "Access Code" field and press the "Submit" button.

PARCC States Access Code: PG6M-835-27WP	SBAC States Access Code: SG6M-648-72WP

Access Code: Please enter your Access Code   **Submit**

Using a smart phone or tablet:
Scan the respective state specific QR code below and follow the instructions.

PARCC States

SBAC States

In the next screen, click on the "New User" button to register your user name and password.

Login
Lumos Common Core Aligned Online Workbooks - 6th Grade Math

If you are a New User, please register.

Login:

Password:

Enter

Subsequent Access:

After you establish your user id and password for subsequent access, simply login with your account information.

What if I buy more than one Lumos Study Program?

Please note that you can use all Online Workbooks with one User ID and Password. If you buy more than one book, you will access them with the same account.

Go back to the **http://www.lumoslearning.com/book** link and enter the access code provided in the second book. In the next screen simply login using your previously created account.

Lumos StepUp® Mobile App FAQ For Students

What is the Lumos StepUp® App?

It is a FREE application you can download onto your Android smart phones, tablets, iPhones, and iPads.

What are the Benefits of the StepUp® App?

This mobile application gives convenient access to Practice Tests, Common Core State Standards, Online Workbooks, and learning resources through your smart phone and tablet computers.

- Eleven Technology enhanced question types in both MATH and ELA
- Sample questions for Arithmetic drills
- Standard specific sample questions
- Instant access to the Common Core State Standards
- Jokes and cartoons to make learning fun!

Do I Need the StepUp® App to Access Online Workbooks?

No, you can access Lumos StepUp® Online Workbooks through a personal computer. The StepUp® app simply enhances your learning experience and allows you to conveniently access StepUp® Online Workbooks and additional resources through your smart phone or tablet.

How can I Download the App?

Visit **lumoslearning.com/a/stepup-app** using your smart phone or tablet and follow the instructions to download the app.

QR Code
for Smart Phone
Or Tablet Users

Lumos SchoolUp™ Mobile App FAQ For Parents and Teachers

What is the Lumos SchoolUp™ App?

It is a FREE App that helps parents and teachers get a wide range of useful information about their school. It can be downloaded onto smartphones and tablets from popular App Stores.

What are the Benefits of the Lumos SchoolUp™ App?

It provides convenient access to

- School "Stickies". A Sticky could be information about an upcoming test, homework, extra curricular activities and other school events. Parents and educators can easily create their own sticky and share with the school community.
- Common Core State Standards.
- Educational blogs.
- StepUp™ student activity reports.

How can I Download the App?

Visit **lumoslearning.com/a/schoolup-app** using your smartphone or tablet and follow the instructions provided to download the App. Alternatively, scan the QR Code provided below using your smartphone or tablet computer.

QR Code
for Smart Phone
Or Tablet Users

Ratios & Proportional Relationships

Expressing Ratios (6.RP.A.1)

1. A school has an enrollment of 600 students. 330 of the students are girls. Express the fraction of students who are boys in simplest terms.

    Ⓐ $\dfrac{12}{20}$

    Ⓑ $\dfrac{11}{20}$

    Ⓒ $\dfrac{9}{20}$

    Ⓓ $\dfrac{13}{20}$

2. In the 14th century, the Sultan of Brunei noticed that his ratio of emeralds to rubies was the same as the ratio of diamonds to pearls. If he had 85 emeralds, 119 rubies, and 45 diamonds, how many pearls did he have?

    Ⓐ 17
    Ⓑ 22
    Ⓒ 58
    Ⓓ 63

3. Mr. Fullingham has 75 geese and 125 turkeys. What is the ratio of the number of geese to the total number of birds in simplest terms?

    Ⓐ 75:200
    Ⓑ 3:8
    Ⓒ 125:200
    Ⓓ 5:8

4. The little league team called the Hawks has 7 brunettes, 5 blonds, and 2 red heads. What is the ratio of redheads to the entire team in simplest terms?

    Ⓐ 2:7
    Ⓑ 2:5
    Ⓒ 2:12
    Ⓓ 1:7

5. The entire little league division that the Hawks belong to has the same ratio of redheads to everyone else. What is the total number of redheads in that division if the total number of players is 126?

    Ⓐ 9
    Ⓑ 14
    Ⓒ 18
    Ⓓ 24

6. Barnaby decided to count the number of ducks and geese flying south for the winter. The first day he counted 175 ducks and 63 geese. What is the ratio of ducks to the total number of birds flying overhead in simplest terms?

    Ⓐ 175:63
    Ⓑ 175:238
    Ⓒ 25:9
    Ⓓ 25:34

7. By the end of migration, Barnaby had counted 4,725 geese. If the ratio of ducks to geese remained the same (175 to 63), how many ducks did he count?

    Ⓐ 13,125
    Ⓑ 17,850
    Ⓒ 10,695
    Ⓓ 14,750

8. Barbara was baking a cake and could not find her tablespoon measure. The recipe calls for $3\frac{1}{3}$ tablespoons. How many teaspoons must Barbara use in order to have the recipe turn out all right?

    Ⓐ 3
    Ⓑ 6
    Ⓒ 9
    Ⓓ 10

9. **The ratio of girls to boys in a grade is 6 to 5. If there are 24 girls in the grade then how many students are there altogether?**

    Ⓐ 14
    Ⓑ 24
    Ⓒ 34
    Ⓓ 44

10. **The ratio of pencils to pens in a box is 3 to 2. If there are 30 pencils and pens altogether, how many pencils are there?**

    Ⓐ 16
    Ⓑ 17
    Ⓒ 18
    Ⓓ 19

Unit Rates (6.RP.A.2)

1. Which is a better price: 5 for $1.00, 4 for 85¢, 2 for 25¢, or 6 for $1.10?

    Ⓐ 5 for $1.00
    Ⓑ 4 for 85¢
    Ⓒ 2 for 25¢
    Ⓓ 6 for $1.10

2. At grocery Store A, 5 cans of baked beans cost $3.45. At grocery Store B, 7 cans of baked beans cost $5.15. At grocery Store C, 4 cans of baked beans cost $2.46. At grocery Store D, 6 cans of baked beans cost $4.00. How much money would you save if you bought 20 cans of baked beans from grocery store C than if you bought 20 cans of baked beans from grocery store A?

    Ⓐ $1.75
    Ⓑ $1.25
    Ⓒ $1.50
    Ⓓ 95¢

3. Beverly drove from Atlantic City to Newark. She drove for 284 miles at a constant speed of 58 mph. How long did it take Beverly?

    Ⓐ 4 hours and 45 minutes
    Ⓑ 4 hours and 54 minutes
    Ⓒ 4 hours and 8 minutes
    Ⓓ 4 hours and 89 minutes

4. Don has two jobs. For Job 1, he earns $7.55 an hour. For Job 2, he earns $8.45 an hour. Last week he worked at the first job for 10 hours and at the second job for 15 hours. What were his average earnings per hour?

    Ⓐ $8.00
    Ⓑ $8.09
    Ⓒ $8.15
    Ⓓ $8.13

5. It took Marjorie 15 minutes to drive from her house to her daughter's school. If the school was 4 miles away from her house, what was her unit rate of speed?

Ⓐ 16 mph
Ⓑ 8 mph
Ⓒ 4 mph
Ⓓ 30 mph

6. The Belmont race track known as "Big Sandy" is 1½ miles long. In 1973, Secretariat won the Belmont Stakes race in 2 minutes and 30 seconds. Assuming he ran on "Big Sandy", what was his unit speed?

Ⓐ 30 mph
Ⓑ 40 mph
Ⓒ 36 mph
Ⓓ 38 mph

7. If 1 pound of chocolate creams at Philadelphia Candies costs $7.52. How much does that candy cost by the ounce?

Ⓐ 48¢ per oz.
Ⓑ 47¢ per oz.
Ⓒ 75.2¢ per oz.
Ⓓ 66¢ per oz.

8. If Carol pays $62.90 to fill the 17-gallon gas tank in her vehicle and she can drive 330 miles on one tank of gas, about how much does she pay per mile to drive her vehicle?

Ⓐ $0.37
Ⓑ $3.70
Ⓒ $0.19
Ⓓ $0.01

9. A 13 ounce box of cereal cost $3.99. What is the unit price per pound?

Ⓐ about $1.23
Ⓑ about $2.66
Ⓒ about $4.30
Ⓓ about $4.91

10. A bottle of perfume costs $26.00 for a 1/2 ounce bottle. What is the price per ounce?

Ⓐ $25.50
Ⓑ $26.50
Ⓒ $52.00
Ⓓ $13.00

Solving Real World Ratio Problems (6.RP.A.3)

1. **How many kilograms are there in 375 grams?**

    Ⓐ 3,750 kg
    Ⓑ 37.5 kg
    Ⓒ 3.75 kg
    Ⓓ 0.375 kg

2. **How many inches are there in 2 yards?**

    Ⓐ 24 in
    Ⓑ 36 in
    Ⓒ 48 in
    Ⓓ 72 in

3. **What is 50% of 120?**

    Ⓐ 50
    Ⓑ 60
    Ⓒ 70
    Ⓓ 55

4. **Michael Jordan is six feet 6 inches tall. How much is that in inches?**

    Ⓐ 66 inches
    Ⓑ 76 inches
    Ⓒ 86 inches
    Ⓓ 78 inches

5. **What is 7.5% in decimal notation?**

    Ⓐ 0.75
    Ⓑ 0.075
    Ⓒ 0.0075
    Ⓓ 7.5

6. **A $60 shirt is on sale for 30% off. How much is the shirt's sale price?**

    Ⓐ $30
    Ⓑ $40
    Ⓒ $18
    Ⓓ $42

7. On Monday, 6 out of every 10 people who entered a store purchased something. If 1,000 people entered the store on Monday, how many people purchased something?

    Ⓐ 6 people
    Ⓑ 60 people
    Ⓒ 600 people
    Ⓓ 610 people

8. If a pair of pants that normally sells for $51.00 is now on sale for $34.00, by what percentage was the price reduced?

    Ⓐ 30%
    Ⓑ 60%
    Ⓒ 33.33%
    Ⓓ 66.67%

9. If Comic Book World is taking 28% off the comic books that normally sell for $4.00, how much money is Kevin saving if he buys 12 comic books during the sale?

    Ⓐ $28
    Ⓑ $12
    Ⓒ $13.44
    Ⓓ $14.58

10. Eric spends 45 minutes getting to work and 45 minutes returning home. What percent of the day does Eric spend commuting?

    Ⓐ 6.25%
    Ⓑ 7.8%
    Ⓒ 5.95%
    Ⓓ 15%

Solving Unit Rate Problems (6.RP.A.3.B)

1. A 12-pack of juice pouches costs $6.00. How much does one juice pouch cost?

    Ⓐ $0.02
    Ⓑ $0.20
    Ⓒ $0.50
    Ⓓ $0.72

2. Eli's can ride his scooter 128 miles on one tank of gas. If the scooter has a 4-gallon gas tank, how far can Eli ride on one gallon of gas?

    Ⓐ 64 miles per gallon
    Ⓑ 32 miles per gallon
    Ⓒ 512 miles per gallon
    Ⓓ 20 miles per gallon

3. Clifton ran 6 miles in 39 minutes. At this rate how fast can Clifton run one mile?

    Ⓐ 13 minutes
    Ⓑ 12 minutes
    Ⓒ 7.2 minutes
    Ⓓ 6 minutes and 30 seconds

4. Brad has swim practice 3 days a week. This week Brad swam a total of 114 laps. At this rate how many laps did Brad swim each day?

    Ⓐ 38 laps
    Ⓑ 42 laps
    Ⓒ 57 laps
    Ⓓ 61 laps

5. Karen bought a total of seven items at five different stores. She began with $65.00 and had $15.00 remaining. Which of the following equation can be used to determine the average cost per item?

    Ⓐ $7x \times 5 = \$50.00$
    Ⓑ $7x = \$75.00$
    Ⓒ $7x + \$15.00 = \$65.00$
    Ⓓ $5x = \$65.00 - \$15.00$

6. Geoff goes to the archery range five days a week. He must pay $1.00 for every ten arrows that he shoots. If he spent $15.00 this week on arrows what is the average number of arrows Geoff shot per day?

Ⓐ 3 arrows
Ⓑ 30 arrows
Ⓒ 45 arrows
Ⓓ 75 arrows

7. Julia made 7 batches of cookies and ate 3 cookies. There were 74 cookies left. Which expression can be used to determine the average number of cookies per batch?

Ⓐ 74÷7
Ⓑ (74+7)÷3
Ⓒ $\dfrac{74 + 3}{7}$
Ⓓ $\dfrac{74}{3} \times 7$

8. Lars delivered 124 papers in 3 hours. How long did it take Lars to deliver one paper?

Ⓐ 1 minute
Ⓑ 1 minute and 27 seconds
Ⓒ 1 minute and 45 seconds
Ⓓ 2 minutes and 3 seconds

9. Mr. and Mrs. Fink met their son Conrad at the beach. Mr. and Mrs. Fink drove 462 miles on 21 gallons of fuel. Conrad drove 456 miles on 12 gallons of fuel. How many more miles per gallon does Conrad's car get than Mr. and Mrs. Fink's car?

Ⓐ 6 mpg
Ⓑ 22 mpg
Ⓒ 16 mpg
Ⓓ 38 mpg

10. Myka bought a box of 30 greeting cards for $4.00. Chuck bought a box of 100 greeting cards for $12.00. Who got the better deal?

Ⓐ Myka got the better deal at about 13 cents per card.
Ⓑ Myka got the better deal at about 7.5 cents per card.
Ⓒ Chuck got the better deal at about 8 cents per card.
Ⓓ Chuck got the better deal at 12 cents per card.

Finding Percent (6.RP.A.3.C)

1. **What is 25% of 24?**

   Ⓐ 5
   Ⓑ 6
   Ⓒ 11
   Ⓓ 17

2. **What is 15% of 60?**

   Ⓐ 9
   Ⓑ 12
   Ⓒ 15
   Ⓓ 25

3. **9 is what percent of 72?**

   Ⓐ 7.2%
   Ⓑ 8%
   Ⓒ 12.5%
   Ⓓ 14%

4. **30% of 190 is what?**

   Ⓐ 45
   Ⓑ 57
   Ⓒ 60
   Ⓓ 63

5. **Daniel has 280 baseball cards. 15% of these are highly collectable. How many baseball cards does Daniel possess that are highly collectable?**

   Ⓐ 15 cards
   Ⓑ 19 cards
   Ⓒ 42 cards
   Ⓓ 47 cards

6. The football team consumed 80% of the water provided at the game. If the team consumed 8-gallons of water, how much water was provided?

   Ⓐ 10 gallons
   Ⓑ 12 gallons
   Ⓒ 15 gallons
   Ⓓ 18.75 gallons

7. Joshua brought 156 of his 678 Legos to Emily's house. What percentage of his Legos did Joshua bring?

   Ⓐ 4%
   Ⓑ 23%
   Ⓒ 30%
   Ⓓ 43%

8. At batting practice Alexis hit 8 balls out of 15 into the outfield. Which equation below can be used to determine the percentage of balls hit into the outfield?

   Ⓐ $\dfrac{15}{8} = \dfrac{x}{100}$

   Ⓑ $\dfrac{15}{100} = \dfrac{x}{8}$

   Ⓒ $8x = (100)(15)$

   Ⓓ $\dfrac{15}{8} = \dfrac{100}{x}$

9. Nikki grows roses, tulips, and carnations. She has 78 flowers of which 32% are roses. Approximately how many roses does Nikki have?

   Ⓐ 18 roses
   Ⓑ 25 roses
   Ⓒ 28 roses
   Ⓓ 41 roses

10. Victor took out 30% of his construction paper. Of this, Paul used 6 sheets, Allison used 8 sheets and Victor and Gayle used the last ten sheets. How many sheets of construction paper did Victor not take out?

   Ⓐ 24 sheets
   Ⓑ 50 sheets
   Ⓒ 56 sheets
   Ⓓ 80 sheets

Measurement Conversion (6.RP.A.3.D)

1. **Owen is 69 inches tall. How tall is Owen in feet?**

    Ⓐ 5.2 feet
    Ⓑ 5.75 feet
    Ⓒ 5.9 feet
    Ⓓ 6 feet

2. **What is 7 gallons 3 quarts expressed as quarts?**

    Ⓐ 4.75 quarts
    Ⓑ 28 quarts
    Ⓒ 29.2 quarts
    Ⓓ 31 quarts

3. **How many centimeters in 3.7 kilometers?**

    Ⓐ 0.000037 cm
    Ⓑ 0.037 cm
    Ⓒ 3700 cm
    Ⓓ 370,000 cm

4. **136 ounces is how many pounds?**

    Ⓐ 6.8 pounds
    Ⓑ 8.5 pounds
    Ⓒ 1088 pounds
    Ⓓ 2,176 pounds

5. **How many ounces in 5 gallons?**

    Ⓐ 128 ounces
    Ⓑ 320 ounces
    Ⓒ 640 ounces
    Ⓓ 1280 ounces

6. Lisa, Susan, and Chris participated in a three-person relay team. Lisa ran 1284 meters, Susan ran 1635 meters and Chris ran 1473 meters. How long was the race in kilometers? Round your answer to the nearest tenth.

Ⓐ 4.0 km
Ⓑ 4.4 km
Ⓒ 43.9 km
Ⓓ 49.0 km

7. Quita recorded the amount of time it took her to complete her chores each week for a month; 1 hour 3 minutes, 1 hour 18 minutes, 55 minutes, and 68 minutes. How many hours did Quita spend doing chores during the month?

Ⓐ 3.8 hours
Ⓑ 4.24 hours
Ⓒ 4.4 hours
Ⓓ 5.7 hours

8. Lamar can run 3 miles in 18 minutes. At this rate, how fast can he run in miles per hour?

Ⓐ 0.9 mph
Ⓑ 1.1 mph
Ⓒ 10 mph
Ⓓ 21 mph

9. A rectangular garden has a width of 67 inches and a length of 92 inches. What is the perimeter of the garden in feet?

Ⓐ 13.25 feet
Ⓑ 26.5 feet
Ⓒ 31.8 feet
Ⓓ 42.8 feet

10. Pat has a pen pal in England. When Pat asked how tall his pen pal was he replied, 1.27 meters. If 1 inch is 2.54 cm, how tall is Pat's pen pal in feet and inches?

Ⓐ 3 feet 11 inches
Ⓑ 4 feet 2 inches
Ⓒ 4 feet 6 inches
Ⓓ 5 feet exactly

End of Ratios & Proportional Relationships

Ratios & Proportional Relationships

Answer Key
&
Detailed Explanations

Expressing Ratios (6.RP.A.1)

Question No.	Answer	Detailed Explanation
1	C	First, to find the proper ratio, subtract the number of girls from the total number of students. The difference is the number of boys. $600-330 = 270$. So, the initial ratio is 270/600. Then, to rewrite a ratio in its simplest terms, divide the numerator and denominator by the Greatest Common Factor (GCF). Here, the GCF is 30. 270 divided by 30 = 9 and 600 divided by 30 = 20, so, the simplest ratio is 9/20.
2	D	First, find the ratio of emeralds to rubies. That ratio is 85/119. To find how many pearls the sultan had, set up a proportion with the ratio of diamonds to pearls: $85/119 = 45/x$ Then, find the cross products of each: $85*x = 119*45$ Simplify: $85x = 5355$ Solve for x by dividing by 85 on both sides: $85x/85 = 5355/85$ $x = 63$
3	B	$75 + 125 = 200$. Therefore, the total number of birds is 200. The ratio of geese to total birds is 75:200. Simplify the ratio by dividing by the GCF of 25, making 3:8.
4	D	There are 14 players in all (7+5+2). The ratio of redheads to the team is 2:14. Divide by the GCF of 2 to simplify the ratio to 1:7
5	C	Set up the proportion: $1/7 = x/126$, cross multiply to get $7x = 126$, then divide by 7 and $x = 18$.
6	D	The total number of birds is $175+63 = 238$. Thus, the ratio of ducks to total birds is 175:238. To find the ratio in simplest terms, divide by the GCF of 7. The ratio in simplest terms is 25:34.

Question No.	Answer	Detailed Explanation
7	A	The ratio of ducks to geese is 175:63. To find how many ducks, set up a proportion of 175/63 = x/4,725. Find the cross products: 175*4,725 = 63*x 826,875 = 63x Divide both sides by 63 x = 13,125
8	D	There are 3 teaspoons to each tablespoon. $3 * 3\frac{1}{3} = 3\frac{10}{3} = 10$ teaspoons.
9	D	To find how many students there are in the grade, set up the proportion 6/5 = 24/x. Notice that you can multiply 6/5 by 4/4 to make the numerator of 24. This makes the equivalent denominator 20. Add 24 + 20 to get the total number of students, or 44.
10	C	If the ratio of pencils to pens is 3/2 then the ratio of pencils to pencils and pens is 3/5. To find the number of pencils in a box with 30 pencils and pens, set up the proportion 3/5 = x/30. Then, multiply the first ratio by 6/6 which will equal 18/30. There are 18 pencils in the box.

Unit Rates (6.RP.A.2)

1	C	5/1.00 = a unit price of $0.20 per piece 4/.85 = a unit price of $0.2125 per piece 2/.25 = a unit price of $0.125 per piece. This is the best price per unit. 6/1.10 = a unit price of $0.183 per piece.
2	C	The unit rate at Store A is $3.45/5=$0.69. 20 cans of beans would be $0.69*20= $13.80 The unit rate at Store C is $2.46/4 = $0.615. 20 cans of beans would be $0.615*20=$12.30. Subtract $13.80−12.30=$1.50
3	B	284 miles divided by 58 miles per hour are how you will find how long it took Beverly to make the trip. (Distance ÷ rate = time) 284/58 ≈ 4.9 hours 0.9 hours = 54 minutes (Multiply 60 by 0.9, because there are 60 minutes in an hour.) 4 hours and 54 minutes is how long it took Beverly to make the trip.

Question No.	Answer	Detailed Explanation
4	B	$7.55 x 10 = $75.55 $8.45 x 15 = $126.75 126.75 + 75.55 = 202.30 202.30/25 = $8.09
5	A	15/4 = 60/x, where 60 equals the number of minutes in an hour. 15 x 4 = 60, so multiply the original ratio 15/4 by 4/4 to get 60/16, where 16 represents the miles per hour (mph) that she traveled.
6	C	Set up a ratio of distance/time. Here, the ratio would be 1.5/2.5 Then, create a proportion 1.5/2.5 = x/60, where 60 represents the number of minutes in an hour. Find the cross products: 1.5*60 = 2.5*x Simplify: 90 = 2.5x Divide each side by 2.5 x = 36
7	B	There are 16 ounces in a pound, so $7.52/16 = 47¢
8	C	To find the cost of gas per mile: $62.90/330 equals about $0.19 per mile. (Note: The capacity of the tank is extra information.)
9	D	$3.99/13 equals about $0.306 per ounce. Since there are 16 oz in a pound, multiply 16 by $0.306..., which equals about $4.91.
10	C	$26.00 ÷ (1/2) = $26.00 x 2 = $52.00 per ounce

Solving Real World Ratio Problems (6.RP.A.3)

Question No.	Answer	Detailed Explanation
1	D	1000 grams/1 kilogram = 375 grams/x kilograms 1000x = 375 Divide each side by 1000 x = 0.375 kilograms
2	D	36 inches equal 1 yard, so 72 inches must equal 2 yards.
3	B	is/of = %/100 so: x/120 = 50/100 100*x = 120*50 100x = 6000 Divide both sides by 100 x = 60
4	D	Since every foot = 12 inches, then 6 feet must equal 72 inches (6*12). Add to that the extra 6 inches to equal 78 inches.

Question No.	Answer	Detailed Explanation
5	B	Divide a percentage by 100 to make an equivalent decimal form 7.5/100 = .075
6	D	is/of = %/100 x/60 = 30/100 x*100 = 60*30 100x = 1800 Divide both sides by 100 x = $18 Subtract $18 from $60. $60−$18 = $42
7	C	6/10 = x/1000 x*10 = 6*1000 10x = 6000 Divide both sides by 10 x = 600
8	C	is/of = %/100 34.00/51.00 = x/100 34.00*100 = 51*x 3400 = 51x Divide both sides by 51 x = 66.67% This is the amount left to pay. 100% − 66.67% = 33.33% This is the amount the shirt was reduced by.
9	C	is/of = %/100 x/$4.00 = 28/100 4*28 = 100*x 112 = 100x Divide both sides by 100 x = $1.12 Then, multiply $1.12 * 12 = $13.44

Question No.	Answer	Detailed Explanation
10	A	is/of = x/100 use hours as your proportional rate 45 minutes + 45 minutes = 90 minutes or 1.5 hours 1.5/24 = x/100 1.5*100 = 24*x 150 = 24x Divide both sides by 24 x = 6.25%

Solving Unit Rate Problems (6.RP.A.3.B)

1	C	Find the unit rate for one juice pouch. $6.00/12 = x/1 6*1=12*x 6 = 12x Divide both sides by 12 x = $0.50 per pouch
2	B	Find the unit rate for one gallon of gas. 128/4 = x/1 128*1=4*x 128 = 4x Divide both sides by 4 x = 32 miles per gallon
3	D	Find the unit rate for one mile. 39/6 = x/1 39*1=6*x 39 = 6x Divide both sides by 6 x = 6.5 or 6 minutes 30 seconds
4	A	Find the unit rate for one day. 114/3 = x/1 114*1=3*x 114 = 3x Divide both sides by 3 x = 38 laps per day

Question No.	Answer	Detailed Explanation
5	C	The cost of the seven items plus $15.00 should equal $65.00. If the average cost per item is x, then 7x is the cost of all seven items. Therefore 7x + $15.00 = $65.00 can be used to find x.
6	B	Find the unit rate for one day. Geoff shot 150 arrows ($15*10) 150/5 days = x/1 150*1=5*x 150 = 5x Divide both sides by 5 x = 30 arrows per day
7	C	Find the total number of cookies and divide by 7. The total number of cookies is 74 + 3 = 77. The number of batches is 7. So the total number of cookies per batch can be found using the expression (74 + 3)/7.
8	B	Find the unit rate for one paper. Change 3 hours to 180 minutes (3*60) 180/124 days = x/1 180*1=124*x 180 = 124x Divide both sides by 124 x = 1.45 minutes or 1 minutes and 27 seconds
9	C	Find the unit rate for both and compare. Fink's: 462 miles/21 gallons = 22 miles per gallon Conrad's: 456 miles/12 gallons = 38 gallons Difference: 38 – 22 = 16 gallons
10	D	Find the unit rate for both and compare. Myka: $4.00/30 cards = $0.13 per card Chuck: $12.00/100 cards = $0.12 per card Chuck paid $0.12 per card and Myka paid $0.13 cents per card so Chuck got the better deal.

Question No.	Answer	Detailed Explanation
Finding Percent (6.RP.A.3.C)		
1	B	is/of = %/100 x/24 = 25/100 x*100 = 24*25 100x = 600 Divide both sides by 100 x = 6
2	A	is/of = %/100 x/60 = 15/100 x*100 = 60*15 100x = 900 Divide both sides by 100 x = 9
3	C	is/of = %/100 9/72 = x/100 9*100 = 72*x 900 = 72x Divide both sides by 72 x = 12.5%
4	B	is/of = %/100 x/190 = 30/100 x*100 = 190*30 100x = 5,700 Divide both sides by 100 x = 57
5	C	is/of = %/100 x/280 = 15/100 x*100 = 280*15 100x = 4,200 Divide both sides by 100 x = 42 cards

Question No.	Answer	Detailed Explanation
6	A	is/of = %/100 8/x = 80/100 8*100 = x*80 800 = 80x Divide both sides by 80 x = 10 gallons
7	B	is/of = %/100 156/678 = x/100 156*100 = 678*x 15,600 = 678x Divide both sides by 678 x = 23%
8	D	is/of = %/100 is the same as of/is = 100/% Is = 8; of = 15 ; % = x Substitute these values to get: 15/8 = 100/x
9	B	is/of = %/100 x/78 = 32/100 x*100 = 78*32 100x = 2496 Divide both sides by 100 x = 24.96 which is approximately 25 roses
10	C	Find the number of sheets used: 6+8+10=24 is/of = %/100 24/x = 30/100 24*100 = x*30 2400 = 30x Divide both sides by 30 x = 80 The number of sheet not taken out is 80−24 = 56 sheets

Measurement Conversion (6.RP.A.3.D)

1	B	There are 12 inches in a foot. 69 inches * (1 foot/12 inches) = 69/12 = 5.75 feet

Question No.	Answer	Detailed Explanation
2	D	There are 4 quarts to a gallon. 7*4 = 28 quarts 28 + 3 = 31 quarts
3	D	There are 100 cm in a meter and 1000 meters in a kilometer. 3.7 km * (1000 m/1 km) * (100 cm/1 m) = 370,000 cm
4	B	There are 16 ounces per pound. 136 ounces * (1 lb/16 oz) = 8.5 pounds
5	C	There are 8 ounces per cup, 2 cups per pint, 2 pints per quart and 4 quarts per gallon. 5 gal * (4 qts/1 gal) * (2 pints/1 qt) * (2 cups/1 pt) * (8 oz/1 cup) = 640 ounces
6	B	Find the total length of the race in meters: 1284 + 1635 + 1473 = 4392 meters There are 1000 meters in 1 kilometer. 4392 m * (1 km/1000m) = 4.392 = 4.4 km
7	C	Find the total number of minutes for the month: 1 h 3 m + 1 h 18 m + 55 m + 68 m = 63 m + 78 m + 55 m + 68 m = 264 minutes. There are 60 minutes in 1 hour. 264 min * (1 hr/60 min) = 4.4 hours
8	C	There are 60 minutes in 1 hour. 3 miles/18 minutes = x miles/60 minutes 3*60 = 18*x 180 = 18x Divide both side by 18 x = 10 miles per hour
9	B	Find the perimeter by adding all four sides of the garden: 67 + 67 + 92 + 92 = 318 in There are 12 inches in a foot. 318 in * (1 foot/12 in) = 26.5 feet
10	B	There are 100 cm in a meter and 2.54 cm in 1 inch. 1.27 meters * (100 cm/1 m) * (1 in/2.54 cm) = 50 inches 50 in / 12 in = 4.17 feet = 4 feet 2 inches

The Number System

Apply and extend previous understandings of multiplication and division to divide fractions by fractions

Division of Fractions (6.NS.A.1)

1. **What is the quotient of 20 divided by one-fourth?**

   Ⓐ 80
   Ⓑ 24
   Ⓒ 5
   Ⓓ 15

2. **Do the following operation:** $1\dfrac{1}{2} \div \dfrac{3}{4} =$

   Ⓐ 4

   Ⓑ $\dfrac{1}{2}$

   Ⓒ $\dfrac{1}{2}$

   Ⓓ 2

3. **Do the following operation:** $3\dfrac{2}{3} \div 2\dfrac{1}{6} =$

   Ⓐ $\dfrac{8}{13}$

   Ⓑ $\dfrac{12}{13}$

   Ⓒ $1\dfrac{5}{13}$

   Ⓓ $1\dfrac{9}{13}$

4. Do the following operation: $2\dfrac{3}{4} \div \dfrac{11}{4} =$

    Ⓐ 1
    Ⓑ 2
    Ⓒ 3
    Ⓓ 4

5. Do the following operation: $\dfrac{7}{8} \div \dfrac{3}{4} =$

    Ⓐ $1\dfrac{1}{6}$

    Ⓑ 2

    Ⓒ $\dfrac{21}{32}$

    Ⓓ $\dfrac{5}{9}$

6. Do the following operation: $6\dfrac{3}{4} \div 1\dfrac{1}{8} =$

    Ⓐ $\dfrac{1}{6}$

    Ⓑ 4

    Ⓒ $5\dfrac{3}{4}$

    Ⓓ 6

7. Complete the following division using mental math.

    7 divided by 1/5

    Ⓐ 35
    Ⓑ 7/5
    Ⓒ 5/7
    Ⓓ 1/35

8. **Complete the following division using mental math.**

    11 divided by 6/6

    Ⓐ 66/66
    Ⓑ 1/11
    Ⓒ 1
    Ⓓ 11

9. **What is the result when a fraction is multiplied by its reciprocal?**

    Ⓐ 1/2
    Ⓑ 10
    Ⓒ 1
    Ⓓ It cannot be determined.

10. **Simplify the following problem. Do not solve.**

    $$\frac{14}{21} \div \frac{28}{7}$$

    Ⓐ $\frac{14}{21} \div \frac{28}{7}$

    Ⓑ $\frac{2}{3} \times \frac{1}{4}$

    Ⓒ 1

    Ⓓ 10

Division of Whole Numbers (6.NS.B.2)

1. A team of 12 players got an award of $1,800 for winning a championship football game. If the captain of the team is allowed to keep $315, how much money would each of the other players get? (Assume they split it equally.)

    Ⓐ $135
    Ⓑ $125
    Ⓒ $150
    Ⓓ $123.75

2. Peter gets a salary of $125 per week. He wants to buy a new television that costs $3,960. If he saves $55 per week, which of the following expressions could he use to figure out how many weeks it will take him to save up enough money to buy the new TV?

    Ⓐ $3,960 ÷ ($125 − $55)
    Ⓑ $3,960 − ($125)($55)
    Ⓒ ($3,960 ÷ $125) ÷ $55
    Ⓓ $3,960 ÷ $55

3. An expert typist typed 9,000 words in two hours. How many words per minute did she type?

    Ⓐ 4,500 words per minute
    Ⓑ 150 words per minute
    Ⓒ 75 words per minute
    Ⓓ 38 words per minute

4. Bethany cut off 18 inches of her hair for "Locks of Love". (Locks of Love is a non profit organization that provides wigs to people who have lost their hair due to chemotherapy.) It took her 3 years to grow it back. How much did her hair grow each month?

    Ⓐ 1 inch
    Ⓑ 2 inches
    Ⓒ 0.25 inches
    Ⓓ 0.5 inches

5. On "Jeopardy," during the month of September, the champions won a total of $694,562. Assuming that there were 22 "Jeopardy" shows in September, what was the average amount won each day by the champions?

    Ⓐ $12,435
    Ⓑ $21,891
    Ⓒ $35,176
    Ⓓ $31,571

6. **A marching band wants to raise $20,000 at its annual fundraiser. If they sell tickets for $20 a piece, how many tickets will they have to sell?**

    Ⓐ 500
    Ⓑ 10,000
    Ⓒ 100
    Ⓓ 1,000

7. **A classroom needs 3,200 paper clips for a project. If there are 200 paper clips in a package, how many packages will they need in all?**

    Ⓐ 160
    Ⓑ 1,600
    Ⓒ 18
    Ⓓ 16

8. **A homebuilder is putting new shelves in each closet he is building. He has 2,592 shelves in his inventory. If each closet needs 108 shelves, how many closets can he build?**

    Ⓐ 2.4
    Ⓑ 108
    Ⓒ 42
    Ⓓ 24

9. **A toy maker needs to make $17,235 per month to meet his costs. Each toy sells for $45. How many toys does he need to sell in order to break even (cover his costs)?**

    Ⓐ 393
    Ⓑ 473
    Ⓒ 373
    Ⓓ 383

10. **A stamp collector collected 4,224 stamps last year. He collected the same amount each month. How many stamps did he collect each month?**

    Ⓐ 422
    Ⓑ 352
    Ⓒ 362
    Ⓓ 252

Operations with Decimals (6.NS.B.3)

1. Three friends went out to lunch together. Ben got a meal that cost $7.25, Frank got a meal that cost $8.16, and Herman got a meal that cost $5.44. If they split the check evenly, how much did they each pay for lunch? (Assume no tax)

    Ⓐ $6.95
    Ⓑ $7.75
    Ⓒ $7.15
    Ⓓ $6.55

2. Which of these is the standard form of twenty and sixty-three thousandths?

    Ⓐ 20.63000
    Ⓑ 20.0063
    Ⓒ 20.63
    Ⓓ 20.063

3. Mr. Zito bought a bicycle for $160. He spent $12.50 on repair charges. If he sold the same bicycle for $215, what would his profit be on the investment?

    Ⓐ $ 147.50
    Ⓑ $ 42.50
    Ⓒ $ 67.50
    Ⓓ $ 55.00

4. A certain book is sold in a paperback version for $4.75 or in a hardcover version for $11.50. If a copy of the book is being purchased for each of the twenty students in Mrs. Jackson's class, how much money altogether would be saved by buying the paperback version, as opposed to the hardcover version?

    Ⓐ $ 155.00
    Ⓑ $ 135.00
    Ⓒ $ 115.00
    Ⓓ $ 145.00

5. **Which of these sets contains all equivalent numbers?**

Ⓐ $\left\{ 0.75 \dfrac{3}{4,} \ 75\%, \ \dfrac{8}{12} \right\}$

Ⓑ $\left\{ 0.100, \ \dfrac{5}{50,} \ 15\%, \ 0.010 \right\}$

Ⓒ $\left\{ \dfrac{3}{8,} \ \begin{array}{c} 35\%, \\ 0.35, \end{array} \ \dfrac{35}{100} \right\}$

Ⓓ $\left\{ \dfrac{9}{25,} \ \begin{array}{c} 36\%, \\ 0.360, \end{array} \ \dfrac{18}{50} \right\}$

6. **Brian is mowing his lawn. He and his family have 7.84 acres. Brian mows 1.29 acres on Monday, 0.85 acres on Tuesday, and 3.63 acres on Thursday. How many acres does Brian have left to mow?**

Ⓐ 2.70
Ⓑ 20.7
Ⓒ 2.07
Ⓓ 0.207

7. **Hector is planting his garden. He makes it 5.8 feet wide and 17.2 feet long. What is the area of Hector's garden?**

Ⓐ 9.976 square feet
Ⓑ 99.76 square feet
Ⓒ 99.76 feet
Ⓓ 997.6 square feet

8. **Chris and 2 of his friends go apple picking. Together they pick a bushel of apples that weighs 28.2 pounds. If Chris and his friends split the bushel of apples evenly among themselves, how many pounds of apples will each person take home?**

Ⓐ 9.4 pounds
Ⓑ 0.94 pounds
Ⓒ 94 pounds
Ⓓ 0.094 pounds

9. Joann and John are hiking over a three-day weekend. They have a total of 67.8 miles that they are planning on hiking. On Friday, they hike half of the miles. On Saturday, they hike another 20 miles. How many miles do they have left to hike on Sunday?

    Ⓐ 1.39 miles
    Ⓑ 31.9 miles
    Ⓒ 13.9 miles
    Ⓓ 139 miles

10. Margaret, Justin, and Leigh are babysitting for the neighbor's children during the summer. Each week they make a total of $72.00, and they split the money evenly. At the end of 4 weeks, how much money did Justin make?

    Ⓐ $130.00
    Ⓑ $144.00
    Ⓒ $72.00
    Ⓓ $96.00

Using Common Factors (6.NS.B.4)

1. **Which of these statements is true of the number 17?**

   Ⓐ It is a factor of 17.
   Ⓑ It is a multiple of 17.
   Ⓒ It is prime.
   Ⓓ All of the above are true.

2. **What are the single digit prime numbers?**

   Ⓐ 2, 3, 5, and 7
   Ⓑ 1, 2, 3, 5, and 7
   Ⓒ 3, 5, and 7
   Ⓓ 1, 3, 5, and 7

3. **Which of the following sets below contains only prime numbers?**

   Ⓐ 7, 11, 49
   Ⓑ 7, 37, 51
   Ⓒ 7, 23, 47
   Ⓓ 2, 29, 93

4. **The product of three numbers is equal to 105. If the first two numbers are 7 and 5, what is the third number?**

   Ⓐ 35
   Ⓑ 7
   Ⓒ 5
   Ⓓ 3

5. **What is the prime factorization of 240?**

   Ⓐ 2 × 2 × 2 × 5
   Ⓑ 2 × 2 × 2 × 2 × 15
   Ⓒ 2 × 2 × 2 × 2 × 5 × 3
   Ⓓ 2 × 2 × 2 × 5 × 3

Name: _____   Date: _____

6. Office A's building complex and the school building next door have the same number of rooms. Office A's building complex has floors with 5 one-room offices on each, and the school building has 11 classrooms on each floor. What is the fewest number of rooms that each building can have?

Ⓐ 16
Ⓑ 6
Ⓒ 55
Ⓓ 5/11

7. How do you know if a number is divisible by 3?

Ⓐ if the ones digit is an even number
Ⓑ if the ones digit ends in a 0 or 5
Ⓒ if the sum of the digits in the number is 3 or a multiple of 3
Ⓓ if the sum of the digits in the number is divisible by 2 and 3

8. Find the Greatest Common Factor (GCF) for 42 and 56.

Ⓐ 21
Ⓑ 14
Ⓒ 7
Ⓓ 2

9. What is the prime factorization of 110?

Ⓐ 10 × 11
Ⓑ 110 × 1
Ⓒ 55 × 2
Ⓓ 2 × 5 × 11

10. What is the LCM of 16 and 24?

Ⓐ 16
Ⓑ 24
Ⓒ 36
Ⓓ 48

Positive and Negative Numbers (6.NS.C.5)

1. Larissa has $4\frac{1}{2}$ cups of flour. She is making cookies using a recipe that calls for $2\frac{3}{4}$ cups of flour. After baking the cookies how much flour will be left?

   Ⓐ $2\frac{3}{4}$ cups

   Ⓑ $2\frac{1}{4}$ cups

   Ⓒ $2\frac{3}{8}$ cups

   Ⓓ $1\frac{3}{4}$ cups

2. The accounting ledger for the high school band showed a balance of $2,123. They purchased new uniforms for a total of $2,400. How much must they deposit into their account in order to prevent it from being overdrawn?

   Ⓐ $382
   Ⓑ $462
   Ⓒ $4,000
   Ⓓ $277

3. Juan is climbing a ladder. He begins on the first rung, climbs up four rungs, but then slides down two rungs. What rung is Juan on?

   Ⓐ 2
   Ⓑ 3
   Ⓒ 4
   Ⓓ 5

4. If last year Julie's net profit was $26,247 after she spent $14,256 on expenses, what was her gross revenue?

   Ⓐ −$40,503
   Ⓑ −$11,991
   Ⓒ $40, 503
   Ⓓ $11,991

5. The amount of snow on the ground increased by 4 inches between 4 p.m. and 6 p.m. If there was 6 inches of snow on the ground at 4 p.m. how many inches were on the ground at 6 p.m.?

    Ⓐ 10 inches
    Ⓑ 14 inches
    Ⓒ 2 inches
    Ⓓ 18 inches

6. The temperature at noon was 20 degrees F. For the next 5 hours it dropped 2 degrees per hour. What was the temperature at 5:00 p.m.?

    Ⓐ 15 degrees
    Ⓑ 10 degrees
    Ⓒ 5 degrees
    Ⓓ 0 degrees

7. Tom enters an elevator that is in the basement, one floor below ground level. He travels three floors down to the parking level and then 4 floors back up. What floor does he end up on?

    Ⓐ ground level
    Ⓑ 2nd floor
    Ⓒ 3rd floor
    Ⓓ basement

8. Which of these numbers would not be found between 6 and 7 on a number line?

    Ⓐ $\dfrac{43}{6}$

    Ⓑ $\dfrac{34}{5}$

    Ⓒ $\dfrac{19}{3}$

    Ⓓ $\dfrac{100}{16}$

9. Stacey lives on a cliff. She lives 652 feet above sea level. When she travels to town, she travels down 491 feet. What elevation is town?

Ⓐ 261 feet above sea level
Ⓑ 161 feet below sea level
Ⓒ 161 feet above sea level
Ⓓ 141 feet above sea level

10. Chris lives in Alaska. When he wakes up, the temperature is −53 degrees. By noon, the temperature has risen 27 degrees. What is the temperature at noon?

Ⓐ −27 degrees
Ⓑ −26 degrees
Ⓒ 26 degrees
Ⓓ −24 degrees

Representing Negative Numbers (6.NS.C.6)

1. **Which of these numbers would be found closest to 0 on a number line?**

    Ⓐ −5

    Ⓑ $-5\dfrac{1}{2}$

    Ⓒ $4\dfrac{1}{2}$

    Ⓓ −4

2. **On a number line, how far apart are the numbers −5.5 and 7.5?**

    Ⓐ 13 units
    Ⓑ 12 units
    Ⓒ 12.5 units
    Ⓓ 2 units

3. **Which numbers does the following number line represent?**

    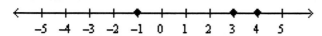

    Ⓐ {−2, 0, 5}
    Ⓑ {−3, −1, 4}
    Ⓒ {−1, 3, 5}
    Ⓓ {−1, 3, 4}

4. **If you go left on a number line, you will** _____

    Ⓐ go in a negative direction
    Ⓑ go in a positive direction
    Ⓒ increase your value
    Ⓓ none of these

5. **Which of the following numbers is NOT between −7 and −4 on the number line?**

    Ⓐ −7.2
    Ⓑ −4.8
    Ⓒ −6
    Ⓓ −5.01

6. **−7.25 is between which two numbers on the number line?**

   Ⓐ −7 and −6
   Ⓑ −5 and −3
   Ⓒ −9 and −10
   Ⓓ −7 and −8

7. **What happens when you start at any number on a number line and add its additive inverse?**

   Ⓐ The number doubles.
   Ⓑ The number halves.
   Ⓒ The sum is zero.
   Ⓓ There is no movement.

8. **Which set of numbers would be found to the left of 4 on the number line?**

   Ⓐ {−1, 4, −5}
   Ⓑ {1, −4, −5}
   Ⓒ {1, 4, −5}
   Ⓓ {1, 4, 5}

9. **On a number line, Bill places a marble on 12. He rolls the marble to the left and it moves 21 spaces and then rolls back to the right 3 spaces. What number does the marble end up on?**

   Ⓐ −21
   Ⓑ −8
   Ⓒ −9
   Ⓓ −6

10. **Which number would be found on the number line between 0 and −1?**

   Ⓐ −1.2
   Ⓑ 1.2
   Ⓒ −0.8
   Ⓓ 0.8

Ordered Pairs (6.NS.C.6.B)

1. **In what Quadrant (I, II, III, IV) does the point (−12, 20) lie?**

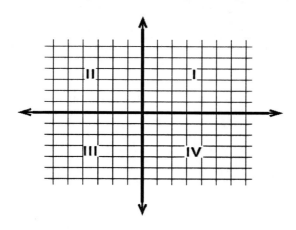

   Ⓐ Quadrant I
   Ⓑ Quadrant II
   Ⓒ Quadrant III
   Ⓓ Quadrant IV

2. **In what Quadrant (I, II, III, IV) does the point (8, −9) lie?**

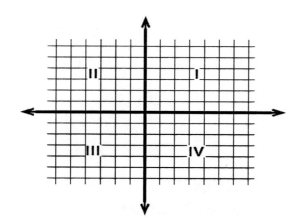

   Ⓐ Quadrant I
   Ⓑ Quadrant II
   Ⓒ Quadrant III
   Ⓓ Quadrant IV

3. **In what Quadrant (I, II, III, IV) does the point (−0.75, −0.25) lie?**

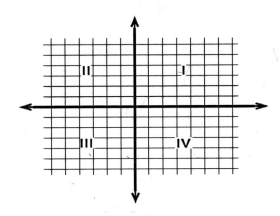

    Ⓐ Quadrant I
    Ⓑ Quadrant II
    Ⓒ Quadrant III
    Ⓓ Quadrant IV

4. **Point A, located at (10,−4), is reflected across the x-axis. What are the coordinates of the reflected point?**

    Ⓐ (−10, −4)
    Ⓑ (10, −4)
    Ⓒ (10, 4)
    Ⓓ (−10, 4)

5. **Point B, located at (6,3), is reflected across the y-axis. What are the coordinates of the reflected point?**

    Ⓐ (6, 3)
    Ⓑ (−6, −3)
    Ⓒ (6, −3)
    Ⓓ (−6, 3)

6. **Point C, located at (2, 9), is reflected across the x-axis called Point C'. Point C' is then reflected across the y-axis called Point C". What are the coordinates of Point C"?**

    Ⓐ (−2, −9)
    Ⓑ (2, 9)
    Ⓒ (2, −9)
    Ⓓ (−2, 9)

7. **Which Quadrant contains the most points in the following list of ordered pairs?**
   (3, 4), (−2,4), (3, −2), (4, −3), (−9, −5), (1, −3), (4, −12), (6, 14), (−5, −11), (−99, −43)

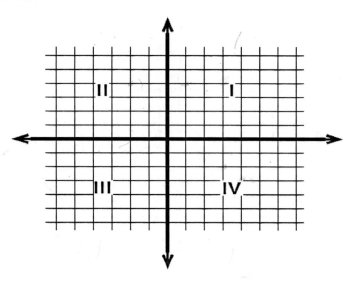

Ⓐ Quadrant I
Ⓑ Quadrant II
Ⓒ Quadrant III
Ⓓ Quadrant IV

8. **The following points have been reflected across the y-axis:** (5, 1), (1, −3), (−5, 3), (−3, 1), (6, −2), (−8, 4), (7, 12). **How many of the reflected points fall in Quadrant II?**

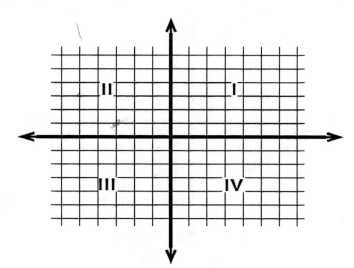

Ⓐ 0
Ⓑ 1
Ⓒ 2
Ⓓ 3

9. The following four points were reflected across the y-axis. A (4, 0), B (0, 0), C (3,0), D(−6,0). Which of the four reflected points (A', B', C', D') is not graphed properly below?

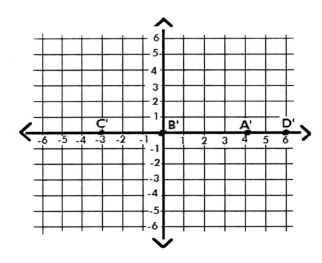

   (A) Point A'
   (B) Point B'
   (C) Point C'
   (D) Point D'

10. A point located at (12, −4) is reflected across the x-axis. In which quadrant (I, II, III, IV) will the reflected point be located?

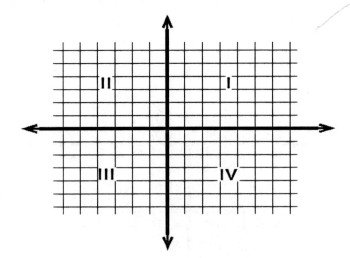

   (A) Quadrant I
   (B) Quadrant II
   (C) Quadrant III
   (D) Quadrant IV

Number Line & Coordinate Plane (6.NS.C.6.C)

1. What number does the dot represent on the number line?

- Ⓐ 12
- Ⓑ 13
- Ⓒ 14
- Ⓓ 15

2. What number does the dot represent on the number line?

- Ⓐ −2
- Ⓑ −1
- Ⓒ 0
- Ⓓ 1

3. What number does the dot represent on the number line?

- Ⓐ 0
- Ⓑ 1
- Ⓒ 4
- Ⓓ 6

4. What number does the dot represent on the number line?

- Ⓐ −5
- Ⓑ −7.5
- Ⓒ −8
- Ⓓ −8.5

5. What number does the dot represent on the number line?

Ⓐ −5
Ⓑ 0
Ⓒ 5
Ⓓ 10

6. Select the point located at (1,−2)

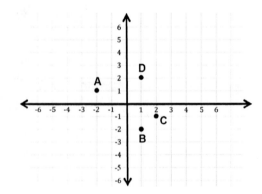

Ⓐ Point A
Ⓑ Point B
Ⓒ Point C
Ⓓ Point D

7. **Select the point located at (–3,5)**

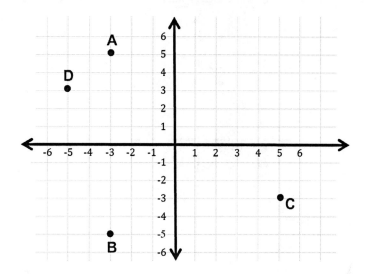

(A) Point A
(B) Point B
(C) Point C
(D) Point D

8. **Select the point located at (–4,–5)**

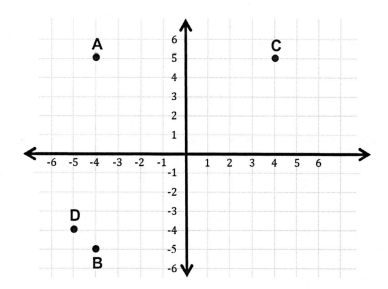

(A) Point A
(B) Point B
(C) Point C
(D) Point D

9. Select the point located at (–1.5, 0.5)

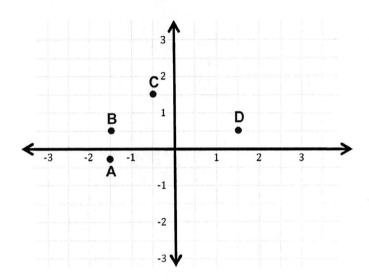

Ⓐ Point A
Ⓑ Point B
Ⓒ Point C
Ⓓ Point D

10. Which of the following points is not graphed below? (3, 4), (–2,4), (3, 2), (–4,–3)

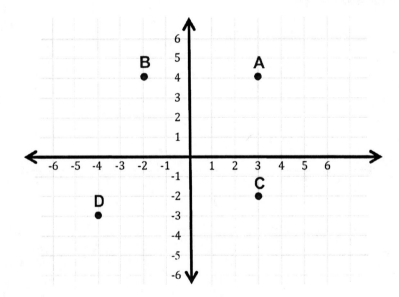

Ⓐ (3, 4)
Ⓑ (–2, 4)
Ⓒ (3, 2)
Ⓓ (–4, 3)

Absolute Value (6.NS.C.7)

1. **Evaluate the following: 17 – |(7)(–3)|**

    Ⓐ 38
    Ⓑ –4
    Ⓒ 4
    Ⓓ 13

2. **Evaluate the following: 16 + |(7)(–3) – 44| – 5**

    Ⓐ 76
    Ⓑ 86
    Ⓒ 34
    Ⓓ 44

3. **Evaluate the following: |15 – 47| + 9 – |(–2)(–4) – 17|**

    Ⓐ 32
    Ⓑ –32
    Ⓒ 50
    Ⓓ 76

4. **Evaluate the following: 18 + 3 |6 – 25| – 11**

    Ⓐ 64
    Ⓑ 100
    Ⓒ 122
    Ⓓ 57

5. **Evaluate the following: 77 – |(–8)(3) + (–10)(4)|**

    Ⓐ 13
    Ⓑ 141
    Ⓒ –141
    Ⓓ –93

6. **Evaluate the following: 21 – |8 – (–5)(7)| – 54**

    Ⓐ –6
    Ⓑ 10
    Ⓒ –118
    Ⓓ –76

7. Evaluate: |9 − 3k| = 3

   Ⓐ k = 0
   Ⓑ k = 2
   Ⓒ k = 2 or k = 4
   Ⓓ k = −4

8. Is the absolute value of a negative integer positive or negative?

   Ⓐ Positive
   Ⓑ Negative
   Ⓒ Neither
   Ⓓ It depends on the magnitude of the number.

9. What is the value of −|24|?

   Ⓐ 1/24
   Ⓑ −1/24
   Ⓒ −24
   Ⓓ 24

10. What is the value of |45 − 75|?

   Ⓐ −115
   Ⓑ 115
   Ⓒ 30
   Ⓓ −30

Rational Numbers in Context (6.NS.C.7.B)

1. **Xavier has a golf score of –7 (We write –7 because it is 7 points below par) and Curtis has a golf score of –12. Who has the higher score?**

    Ⓐ Xavier has the higher score.
    Ⓑ Curtis has the higher score.
    Ⓒ Because both scores are below par, neither one has the higher score.
    Ⓓ We cannot tell who has the higher score because we do not know what par is.

2. **Kelly has read $\frac{5}{6}$ of a book. Helen has read $\frac{9}{12}$ of the same book. Who has read more of the book?**

    Ⓐ $\frac{5}{6}$ is more than $\frac{9}{12}$ so Kelly has read more.

    Ⓑ $\frac{5}{6}$ is less than $\frac{9}{12}$ so Helen has read more.

    Ⓒ $\frac{5}{6}$ is the same as $\frac{9}{12}$ so both Kelly and Helen have read the same amount.

    Ⓓ We cannot tell who has read more because the fractions have different denominators.

3. **The record low temperature for NY is –52°F. The record low temperature for Alaska is –80°F. Which of the following inequalities accurately compares these two temperatures?**

    Ⓐ –52° F < –80° F
    Ⓑ –80° F > –52° F
    Ⓒ –52° F= –80° F
    Ⓓ –52° F > –80° F

4. **A cake recipe calls for $1\frac{3}{4}$ cups of soy flour, $\frac{20}{8}$ cups of rice flour and 1.6 cups of wheat flour. Which of the following inequalities compares these three quantities accurately?**

    Ⓐ $1.6 < \frac{20}{8} < 1\frac{3}{4}$

    Ⓑ $\frac{20}{8} > 1\frac{3}{4} > 1.6$

    Ⓒ $\frac{20}{8} > 1\frac{3}{4} < 1.6$

    Ⓓ $1\frac{3}{4} > 1.6 > \frac{20}{8}$

5. Doug and Sissy went scuba diving. Doug descended to –143 feet and Sissy descended to –134 feet. Who dove deeper?

Ⓐ –134 > –143, so Sissy dove deeper.
Ⓑ –134 < –143, so Doug dove deeper.
Ⓒ –134 = –143, so neither one dove deeper as they descended the same amount.
Ⓓ |–143|>|–134|, so Doug dove deeper.

6. At the annual town festival $\frac{4}{15}$ of the vendors sold outdoor items, 0.4 sold clothing or indoor items and $\frac{1}{3}$ sold food. Which of the following inequalities compares these three quantities accurately?

Ⓐ $0.4 < \frac{4}{15} < \frac{1}{3}$

Ⓑ $\frac{4}{15} > 0.4 > \frac{1}{3}$

Ⓒ $0.4 > \frac{1}{3} > \frac{4}{15}$

Ⓓ $\frac{1}{3} < \frac{4}{15} < 0.4$

7. A school of fish (S1) is spotted in the ocean at 15 feet below sea level. A second school (S2) of fish is spotted at $\frac{33}{3}$ feet below sea level. A third school (S3) of fish is spotted 11.5 feet below sea level. Order these numbers from deepest to shallowest. Note: The symbol > means deeper and < means shallower

Ⓐ S1 < S3 < S2
Ⓑ S3 < S2 < S1
Ⓒ S1 > S3 > S2
Ⓓ S3 < S1 < S2

8. During the first snowfall of the year, Henderson recorded the snow fall each day. The first day $\frac{4}{5}$ of a foot fell. On the second day $\frac{5}{7}$ of a foot fell. On which day did more snow fall?

Ⓐ $\frac{4}{5} > \frac{5}{7}$, so more snow fell on the first day.

Ⓑ $\frac{4}{5} < \frac{5}{7}$, so more snow fell on the second day.

Ⓒ $\frac{4}{5} = \frac{5}{7}$, so the same amount of snow fell on both days.

Ⓓ We cannot tell from this information because the fractions have different denominators.

9. Molly has $365 in her savings account. She withdraws $415. Bill has a savings account balance of −$45. Which of the following statements is correct?

Ⓐ Both Molly and Bill owe the bank the same amount.
Ⓑ Molly owes the bank more than Bill.
Ⓒ Bill owes the bank more than Molly.
Ⓓ Neither Molly or Bill owe the bank any money.

10. Jeremiah and Farley each bought 5 boxes of energy bars. Within a week Jeremiah eats 2 $\frac{5}{6}$ boxes and Farley eats 1 $\frac{15}{9}$ boxes. Who has more left?

Ⓐ Jeremiah has more left.
Ⓑ Farley has more left.
Ⓒ They both have the same amount left.
Ⓓ We cannot tell from this information because the mixed numbers have different denominators.

Interpreting Absolute Value (6.NS.C.7.C)

1. **What is the absolute value of the number represented by the red dot plotted below?**

    -7   0   8

    (A) −3
    (B) 1
    (C) 3
    (D) 4

2. **What is the absolute value of the number represented by the red dot plotted below?**

    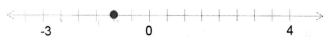

    -3   0   4

    (A) −2
    (B) −1
    (C) 1
    (D) 2

3. **Which symbol will make the following a true statement? |−27| _____ 19**

    (A) >
    (B) <
    (C) =
    (D) ≤

4. **Which symbol will make the following a true statement? −4 _____ −|−6|**

    (A) >
    (B) <
    (C) =
    (D) ≤

5. **Which of the following numbers will make both inequalities true? −12 < x; |x| < 4**

    (A) x = −5
    (B) x = −1
    (C) x = −7
    (D) x = −4

6. Ruth, who lives in Florida at an elevation of 30 meters, goes on a vacation to the Grand Cayman Islands, at an elevation of 24 meters, to go scuba diving at an elevation of −30 meters. Which elevation has the greatest absolute value?

Ⓐ 30 meters
Ⓑ 24 meters
Ⓒ −30 meters
Ⓓ Both 30 meters and −30 meters have the greater absolute value.

7. Connie, Julie, and Shelley's parents have encouraged them to save their money. Connie has an account balance of $215, Julie has −$100, and Shelley has −$250. Which inequality accurately represents the relative absolute values of each account.

Ⓐ |$215|<|−$100|<|−$250|
Ⓑ |−$250|<|−$100|<|$215|
Ⓒ |−$100|>|$215|>|−$250|
Ⓓ |−$100|<|$215|<|−$250|

8. The buoys on a certain lake mark the distance in meters from a center buoy. All buoys directly south are given negative numbers and all buoys directly north are given positive numbers. Betsy is located at buoy −6.2 and her brother Luis is located at buoy 6.5. Based on this information, which one of the following statements is not true.

Ⓐ Betsy is closer to the center buoy than Luis
Ⓑ Luis is 6.5 meters from the center buoy.
Ⓒ Luis is 0.2 meters farther from the center buoy than Betsy.
Ⓓ Betsy is 12.7 meters from Luis.

9. Over the last three months Ophelia, Aaron, Nathan, and Rebecca recorded their weight. The table shows their initial and final weights.

Name	Initial Weight, lbs	Final Weight, lbs
Ophelia	145	157
Aaron	178	163
Nathan	205	217
Rebecca	136	128

Which state has the broadest range of temperature extremes?

Ⓐ Ophelia
Ⓑ Aaron
Ⓒ Nathan
Ⓓ Rebecca

10. The table below records the lowest and highest temperatures for four states.

State	Lowest Temp, °F	Highest Temp, °F
New York	−52	108
Texas	−23	120
Florida	−2	109
North Carolina	−34	110

Whose weight had the least absolute change?

Ⓐ New York
Ⓑ Texas
Ⓒ Florida
Ⓓ North Carolina

Comparisons of Absolute Value (6.NS.C.7.D)

1. **Which account balance represents the greatest debt?**
   $20, –$45, –$5, $10

   Ⓐ $20
   Ⓑ –$45
   Ⓒ –$5
   Ⓓ $10

2. **Which of the following is the warmest temperature?**
   5°F above zero, 6°F below zero, 10°F below zero, 2°F above zero

   Ⓐ 5°F above zero
   Ⓑ 6°F below zero
   Ⓒ 10°F below zero
   Ⓓ 2°F above zero

3. **Anneliese spent $58.00 on music items and paid $35.00 on a lay-away item. If she has –142.00 left in her account, how much did she start with?**

   Ⓐ $–49.00
   Ⓑ $93.00
   Ⓒ $135.00
   Ⓓ –$235.00

4. **The Murphys began their summer trip from home at an elevation of 439 feet. They drove to the mountains and climbed to an elevation of 1839 feet. After their visit on the top of the mountain they drove down the mountainside 527 feet where they stopped for lunch. At the end of the day they spent the night at a campground at an elevation 264 feet higher than the restaurant. What is the difference in elevation between the campground and home?**

   Ⓐ 1048 feet
   Ⓑ 1137 feet
   Ⓒ 1576 feet
   Ⓓ 2012 feet

5. Before Tonya went shopping she had $165.00 in her account. She returned a jacket that cost $46.50, bought 2 pairs of socks for $5.99 each, and went to lunch and a movie for $28.00. What is Tonya's account balance now?

    Ⓐ $125.02
    Ⓑ $158.48
    Ⓒ $171.52
    Ⓓ $177.51

6. The Casey quadruplets live in four different states. Dominik lives in Nantucket, Massachusetts at an elevation of 28 feet; Denzel lives in New Orleans, Louisiana at an elevation of 5.3 feet below sea level; Kaila lives in California near Death Valley at an elevation of 7 feet below sea level; and Malik lives in Nome, Alaska at an elevation of 20 feet. Who lives at the lowest elevation?

    Ⓐ Dominik
    Ⓑ Denzel
    Ⓒ Kaila
    Ⓓ Malik

7. One day in January the Casey quadruplets compared their location temperature.

Location	Temperature, °C
Nome, Alaska	14.9 below zero
Nantucket, Massachusetts	5.1 below zero
New Orleans, Louisiana	6 above zero
Death Valley, California	15 above zero

    Which location has the warmest temperature?

    Ⓐ Nome, Alaska
    Ⓑ Nantucket, Massachusetts
    Ⓒ New Orleans, Louisiana
    Ⓓ Death Valley, California

8. Sato, her brother Ichiro, and two friends, Aran and Mio, went to a festival. Before they could board any ride they had to be taller than the wooden height checker at each ride. At one ride Sato was 4 inches taller, Ichiro was 2 inches shorter, Aran was 6 inches taller and Mio was 2.5 inches shorter than the wooden height checker. Who is the shortest person?

    Ⓐ Sato
    Ⓑ Ichiro
    Ⓒ Aran
    Ⓓ Mio

9. Leary went to a sports store with $60.00 and bought a sweat shirt, athletic tape and baseball socks. If he received $3.72 in change, how much did his purchases cost?

 Ⓐ $63.72
 Ⓑ $61.18
 Ⓒ $56.28
 Ⓓ $47.28

10. Reilley owes $75.38 for his phone and $35.00 in dues. He expects to receive a credit of $26.16 for a wrong charge. If he currently has $629.00 in his account, what will be his account balance after all debts and credits are completed?

 Ⓐ $713.22
 Ⓑ $544.78
 Ⓒ $518.62
 Ⓓ $492.46

Coordinate Plane (6.NS.C.8)

1. **Ricky and Becca are going hiking. Below is the map that they are using. They start out at (−3.6, −2.6). They hike four units to the east and six units to the north. What are the coordinates of their new location? (Note: North is up on this map.)**

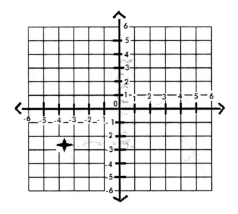

   Ⓐ (0.6, 3.6)
   Ⓑ (0.4, 3.4)
   Ⓒ (0.4, 3.6)
   Ⓓ (1.6, 3.6)

2. **The absolute value of the coordinates are (5, 8). What are the coordinates in Quadrant II?**

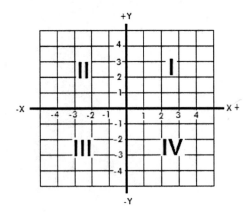

   Ⓐ (−5, −8)
   Ⓑ (5, −8)
   Ⓒ (5, 8)
   Ⓓ (−5, 8)

3. **What is the absolute value of the coordinates shown on the graph?**

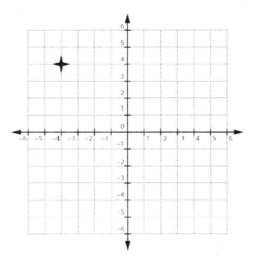

Ⓐ (−4, 4)
Ⓑ (4, 4)
Ⓒ (4, −4)
Ⓓ (−4, −4)

4. **Greg is in the jungle taking pictures of wildlife. He has to walk 5 units north and 3 units west to get back to his village. What are the coordinates of Greg's village? (Note: North is up on this map.)**

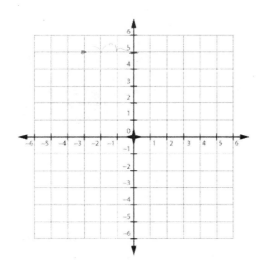

Ⓐ (−3, 5)
Ⓑ (3, 5)
Ⓒ (−3, −5)
Ⓓ (3, −5)

5. **How many units does Yolanda need to walk from point C to point D? (Note: North is up on this map.)**

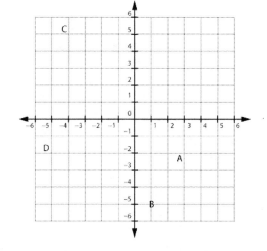

Ⓐ 1 unit west and 7 units north
Ⓑ 1 unit east and 7 units south
Ⓒ 1 unit west and 7 units south
Ⓓ 7 units west and 1 unit south

6. **What is the absolute value of Point D's coordinates?**

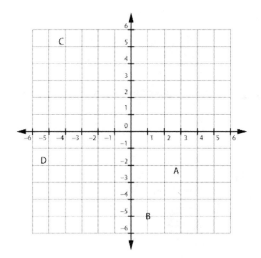

Ⓐ (−5, 2)
Ⓑ (5, −2)
Ⓒ (−5, −2)
Ⓓ (5, 2)

7. **There is a set of coordinates in Quadrant 4. The absolute value of x = 7. the absolute value of y = 3. What are the coordinates?**

Ⓐ (−7, −3)
Ⓑ (−3, −7)
Ⓒ (7, −3)
Ⓓ (−7, 3)

8. There is a set of coordinates in Quadrant 2. The absolute value of x = 3. the absolute value of y = 9. What are the coordinates?

Ⓐ (−9, −3)
Ⓑ (−3, −9)
Ⓒ (9, −3)
Ⓓ (−3, 9)

9. The graph below shows the top of a mountain. If the value y = 0 is at sea level and the y-axis is measuring altitude, how far below sea level is the point (0, −6)?

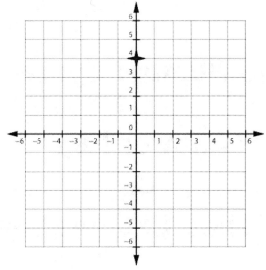

Ⓐ 10 units below sea level
Ⓑ 6 units above sea level
Ⓒ 6 units below sea level
Ⓓ 10 units above sea level

10. The coordinates (−2, 8) are in Quadrant II. If the absolute values of x and y were to stay the same, but the point was moved to Quadrant IV, what would the coordinates be?

Ⓐ (8, −2)
Ⓑ (−8, 2)
Ⓒ (−2, −8)
Ⓓ (2, −8)

End of The Number System

The Number System

Answer Key
&
Detailed Explanations

Division of Fractions (6.NS.A.1)

Question No.	Answer	Detailed Explanation
1	A	The original problem is: $$\frac{20}{1} \div \frac{1}{4} =$$ To divide fractions, you must Keep (the first fraction), Change (the division to multiplication), and Flip (the second fraction, or, take the reciprocal). 20/1 X 4/1 = 80/1 80
2	D	The original problem is: $$1\frac{1}{2} \div \frac{3}{4} =$$ First, convert the mixed number to an improper fraction (bottom times the side plus the top = (2*1)+1 = 3/2 To divide fractions, you must Keep (the first fraction), Change (the division to multiplication), and Flip (the second fraction, or, take the reciprocal). $$\frac{3}{2} \times \frac{4}{3} = \frac{12}{6}$$ Simplify by factoring out the GCF of 6. The answer is 2/1 or 2
3	D	The original problem is: 3 2/3 ÷ 2 1/6 = First, convert the first mixed number to an improper fraction (bottom times the side plus the top = (3*3)+2= 11/3. Then, find the improper fraction of the second mixed number (bottom times the side plus the top = (2*6)+1 = 13/6 To divide fractions, you must Keep (the first fraction), Change (the division to multiplication), and Flip (the second fraction, or, take the reciprocal). $$\frac{11}{3} \times \frac{6}{13} = \frac{66}{39}$$ Simplify by factoring out the GCF of 3. The answer is 22/13 Divide 22/13 to get a mixed number: The answer is 1 9/13

Question No.	Answer	Detailed Explanation
4	A	The original problem is: 2 3/4 ÷ 4/11= First, convert the mixed number to an improper fraction (bottom times the side plus the top = (4*2)+3= 11/4 To divide fractions, you must Keep (the first fraction), Change (the division to multiplication), and Flip (the second fraction, or, take the reciprocal). $$\frac{11}{4} \times \frac{4}{11} = \frac{44}{44} = 1$$
5	A	The original problem is: 7/8 ÷ 3/4 = To divide fractions, you must Keep (the first fraction), Change (the division to multiplication), and Flip (the second fraction, or, take the reciprocal). $$\frac{7}{8} \times \frac{4}{3} = \frac{28}{24}$$ Simplify by factoring out the GCF of 4. The answer is 7/6 Divide 7/6 to get a mixed number: The answer is 1 1/6
6	D	The original problem is: 6 3/4 ÷ 1 1/8 = First, find the improper fraction of the first mixed number (bottom times the side plus the top = (4*6)+3= 27/4 Then, find the improper fraction of the second mixed number (bottom times the side plus the top = (8*1)+1 = 9/8 To divide the fractions, you must Keep (the first fraction), Change (the division to multiplication), and Flip (the second fraction, or, take the reciprocal). $$\frac{27}{4} \times \frac{8}{9}$$ Cross factor out the GCF of 2 from 4 and 8 Cross factor out the GCF of 9 from 9 and 27 $$\frac{3}{1} \times \frac{2}{1} = \frac{6}{1} = 6$$

Question No.	Answer	Detailed Explanation
7	A	To divide the fractions, you must Keep (the first fraction), Change (the division to multiplication), and Flip (the second fraction, or, take the reciprocal). The second fraction then reads 5/1. Because 5/1 is the same as 5, the problem simplifies to 7 X 5 = 35.
8	D	Since 6/6 is equal to 1, the problem simplifies to 11 divided by 1. The answer is 11.
9	C	When any fraction is multiplied by its reciprocal, the cross numerators and denominators will always factor to 1.
10	B	$$\frac{14}{21} \div \frac{28}{7}$$ becomes $$\frac{14}{21} \times \frac{7}{28}$$ after Keep−Change−Flip. Then, cross factor out a GCF of 14 from the 14 and 28, and a GCF of 7 from the 7 and 28. The simplified problem becomes: $$\frac{2}{3} \times \frac{1}{4}$$

Division of whole Numbers (6.NS.B.2)

Question No.	Answer	Detailed Explanation
1	A	First, find the difference of $1,800 − $315 = $1,485 Then, divide $1,485/11 = $135
2	D	To find the answer, divide $3,960 ÷ 55. (Note: The $125 is unnecessary information.)
3	C	1 hour = 60 minutes, so 2 hours = 120 minutes So, 9,000 ÷ 120 = 75
4	D	1 year = 12 months, so 3 years = 36 months 18 ÷ 36 = .5 inches per month
5	D	Use long division. $694,562 ÷ 22 = $31,571
6	D	Use place value concepts or long division to solve. $20,000 ÷ $20 = 1,000
7	D	Use long division or related multiplication facts. 3,200 ÷ 200 = 16
8	D	Use long division or related multiplication facts. 2,592 ÷ 108 = 24
9	D	Use long division. $17,235 ÷ $45 = 383
10	B	Use the standard long division algorithm. 4,224 ÷ 12 = 352

Question No.	Answer	Detailed Explanation
		**Operations with Decimals (6.NS.B.3)**
1	A	$7.25 + $8.16 + $5.44 = $20.85 $20.85/3 = $6.95
2	D	The standard form would be written 20.063 since the whole number part is 20 and the decimal part is written .063 (Sixty-three thousandths).
3	B	$160 + $12.50 = $172.50 (His total investment) $215.00 − $172.50 = $42.50
4	B	$11.50*20 = $230 $4.75*20 = $95 $230 − $95 = $135 $135.00 would be saved by purchasing the paperback book
5	D	The last set is the only set of numbers that are all equivalent. These numbers all have a decimal equivalent of 0.36.
6	C	1.29 + 0.85 + 3.63= 5.77 7.84 − 5.77 = 2.07 acres left to mow
7	B	Multiply the length by width to find the area. 5.8 * 17.2 Multiply the numbers and find the product. Then, count the decimal places in the factors. There is one decimal place for each factor or a total of two decimal places. Count two spaces from the right and put the decimal point in your answer. 5.8 * 17.2 = 99.76
8	A	The bushel of apples weighs 28.2 pounds. There are three people to split the apples. Solve by dividing the total weight of the apples by three. 28.2/3 = 9.4 pounds each
9	C	To figure out how far Joann and John hiked on Friday divide 67.8 by 2. 67.8 ÷ 2 = 33.9. Subtract 20 from 33.9 to find out how much farther they have to travel. 33.9 − 20 =13.9 miles
10	D	To find how much money each person earns per week, divide the total earned by 3. $72.00 ÷ 3 = $24.00 Then, to find how much one person makes over a four-week period, multiply by 4. $24.00 * 4= $96.00

Question No.	Answer	Detailed Explanation

Using Common Factors (6.NS.B.4)

Question No.	Answer	Detailed Explanation
1	D	All statements are true. A number is always a factor and a multiple of itself. It is prime because the only two factors are 1 and 17.
2	A	Other than 1, which is not prime by definition, 2, 3, 5, and 7 are the only single-digit numbers that can be divided by only themselves and 1.
3	C	A prime number is a whole number (greater than 1) which is divisible by only 1 and itself. The set {7, 23, 47} contains three numbers which fit the definition stated above. Each number is divisible by only 1 and itself.
4	D	$(7*5) * x = 105$ $35x = 105$ $x = 3$
5	C	The prime factorization of a number are the prime numbers that when multiplied together, give a product of the starting number. $240 = 120 \times 2$ $120 = 60 \times 2$ $60 = 30 \times 2$ $30 = 15 \times 2$ $15 = 5 \times 3$ $2 \times 2 \times 2 \times 2 \times 3 \times 5 = 240$
6	C	Find the least multiple common to both numbers, starting with the greater number: 11: 11, 22, 33, 44, 55, 66, 77, 88... 5: 5, 10, 15, 20, 25, 30, 35, 40, 45, 50, 55 Stop when you get to the lowest multiple that is common to both. 55 is the LCM (lowest common multiple) of these two numbers, so each building could have 55 rooms.
7	C	When the sum of the digits in the number equals three or a multiple of 3, the number is divisible by 3. For example, 375 is divisible by 3 because $3 + 7 + 5 = 15$ Because 15 is a multiple of 3, the number is divisible by 3.
8	B	42: 1, 2, 3, 6, 7, 14, 21, 42 56: 1, 2, 4, 7, 8, 14, 28, 56

Question No.	Answer	Detailed Explanation
9	D	Prime factorization is when a number is factored as far as possible into its prime factors. Start by factoring a number into two factors: 110 can be factored into two factors of 11 X 10. 11 is a prime number, so that is the first prime factor. 10 is composite so it can be factored into factors of 2 and 5. 2 and 5 are both prime numbers, so those are the remaining prime factors. Thus, the prime factorization of 110 is 2 X 5 X 11
10	D	To find the LCM, list the first few multiples of each number: 16: 16, 32, 48, 64 24: 24, 48, 72 48 is the least multiple that is common to both lists.

Positive and Negative Numbers (6.NS.C.5)

1	D	4.25 -2.75 ——— 1.75 $1.75 = 1\frac{3}{4}$
2	D	$\$2{,}123 - 2{,}400 = -\$277$ They must make a deposit of \$277 in order to prevent the account from being overdrawn.
3	B	$1 + 4 - 2 = 3$
4	C	$\$26{,}247 + \$14{,}256 = \$40{,}503$ The net profit and the expenses must be added together to get the gross revenue.
5	A	$6 + (+4) = 10$ inches
6	B	The change in temperature can be expressed as $5*(-2)$, or -10 degrees $20 + (-10) = 10$ degree F.
7	A	Call the basement 0, since the ground level would be considered the 1st floor. Tom's motion on the elevator can be described as $0 - 3 + 4 = 1$. He ends up on the 1st floor (ground level).
8	A	$\frac{43}{6}$ simplifies to 7 and 1/6, which is larger than 7. All other fractions are greater than 6 but less than 7.

Question No.	Answer	Detailed Explanation
9	C	652 − 491 = 161 feet Because Stacey started out above sea level and does not go below zero, you know the town is 161 feet above sea level.
10	B	−53 + 27 = −26

Representing Negative Numbers (6.NS.C.6)

1	D	−4 is only 4 units away from 0 on a number line. All of the other numbers are farther away.
2	A	If you count the units between −5.5 and 7.5, you will count 13 units between them.
3	D

The number line above shows −1, 3, and 4.
All other answer choices are not represented on this number line. |
4	A	Moving left on the number line indicates moving in a negative direction. Moving right on the number line indicates moving in a positive direction.
5	A	−7.2 is less than −7 so it would not be found between −7 and −4.
6	D	−7.25 is less than −7 but greater than −8 so it would be the number found between −7 and −8.
7	C	Example: 4 + (−4) = 0
8	B	{1, −4, −5} is the only set which would fall to the left of 4 on a number line. Any number larger than 4 would fall to the right on a number line and would not fit the criteria.
9	D	+12 + (−21) + 3 = −6
10	C	The only number that would fall between 0 and −1 would be −0.8, as −1 < −0.8 < 0

Ordered Pairs (6.NS.C.6.B)

1	B	Quadrant II – In this quadrant the x-coordinate is negative and the y-coordinate is positive.
2	D	Quadrant IV – In this quadrant the x-coordinate is positive and the y-coordinate is negative.
3	C	Quadrant III – In this quadrant both the x-coordinate and y-coordinate are negative.

Question No.	Answer	Detailed Explanation
4	C	When a point is reflected across the x-axis the x-coordinate remains the same and the y-coordinate changes signs. (10, −4) → (10, 4)
5	D	When a point is reflected across the y-axis the y-coordinate remains the same and the x-coordinate changes signs. (6, 3) → (−6, 3)
6	A	When a point is reflected across the x-axis the x-coordinate remains the same and the y-coordinate changes signs. C (2, 9) → C' (2, −9). When a point is reflected across the y-axis the y-coordinate remains the same and the x-coordinate changes signs. C' (2, −9) → C'' (−2, −9)
7	D	Quadrant I: (3,4); (6,14) 2 Quadrant II: (−2,4) 1 Quadrant III: (−9,−5), (−5,−11); (−99, −43) 3 Quadrant IV: (3,−2); (4,−3), (1,−3); (4,−12) 4 Quadrant IV has the most points.
8	C	In Quadrant II the x-coordinate is negative and the y-coordinate is positive. When a point is reflect across the y-axis the y-coordinate remains the same and the x-coordinate changes sign. Therefore points with a positive x-coordinate and a positive y-coordinate will fall in Quadrant II when reflected across the y-axis. Therefore two of the points will fall in Quandrant II when reflected across the y-axis: (5,1) and (7,12).
9	A	Point A' (−4,0) was not reflected properly. It should have been graphed four units to the left of the origin. (Note: when the point (0,0) is reflected across the y-axis it remains in the same position.)
10	A	When a point is reflected across the x-axis the x-coordinate remains the same and the y-coordinate changes signs. (12, −4) → (12, 4) The point (12,4) is located in Quadrant I.

Number Line & Coordinate Plane (6.NS.C.6.C)

1	C	Each tick is 1 unit. Therefore the red dot represents the number 14.
2	C	Each tick is 1 unit. Therefore the red dot represents the number 0.
3	C	Each tick is 2 units. Therefore the red dot represents the number 4.
4	B	Each tick is 0.5 units. Therefore the red dot represents the number −7.5.

Question No.	Answer	Detailed Explanation
5	C	Each tick is 10 units. Therefore the red dot represents the number 5.
6	B	An ordered pair is written in the format (x,y) where x is the first number and y is the second number. The point $(1,-2)$ is located 1 unit to the right (x-axis) and 2 units down (y-axis) from the origin. This is represented by Point B.
7	A	An ordered pair is written in the format (x,y) where x is the first number and y is the second number. The point $(-3,5)$ is located 3 units to the left (x-axis) and 5 units up (y-axis) from the origin. This is represented by Point A.
8	B	An ordered pair is written in the format (x,y) where x is the first number and y is the second number. The point $(-4, -5)$ is located 4 units to the left (x-axis) and 5 units down (y-axis) from the origin. This is represented by Point B.
9	B	An ordered pair is written in the format (x,y) where x is the first number and y is the second number. The point $(-1.5, 0.5)$ is located 1.5 units to the left (x-axis) and 0.5 units up (y-axis) from the origin. This is represented by Point B.
10	C	(3,4) = Point A $(-2,4)$ = Point B (3,2) is not plotted $(-4,3)$ = Point D

Absolute Value (6.NS.C.7)

1	B	$17 - |(7)(-3)|$ $= 17 - |-21|$ $= 17 - 21$ $= -4$ Note: Absolute value (the value of $|x|$) is the value of a number without regard to its sign.

Question No.	Answer	Detailed Explanation
2	A	$16 + \lvert(7)(-3) - 44\rvert - 5$ $= 16 + \lvert-21 - 44\rvert - 5$ $= 16 + \lvert-21 + (-44)\rvert - 5$ $= 16 + \lvert-65\rvert - 5$ $= 16 + 65 - 5$ $= 81 - 5$ $= 76$ Note: Absolute value (the value of $\lvert x\rvert$) is the value of a number without regard to its sign.
3	A	$\lvert 15 - 47\rvert + 9 - \lvert(-2)(-4) - 17\rvert$ $= \lvert-32\rvert + 9 - \lvert 8 - 17\rvert$ $= 32 + 9 - \lvert-9\rvert$ $= 32 + 9 - 9$ $= 41-9$ $= 32$ Note: Absolute value (the value of $\lvert x\rvert$) is the value of a number without regard to its sign.
4	A	$18 + 3\lvert 6 - 25\rvert - 11$ $= 18 + 3\lvert-19\rvert - 11$ $= 18 + 3(19) - 11$ $= 18 + 57 - 11$ $= 18 + 57 - 11$ $= 75 - 11$ $= 64$ Note: Absolute value (the value of $\lvert x\rvert$) is the value of a number without regard to its sign.
5	A	$77 - \lvert(-8)(3) + (-10)(4)\rvert$ $= 77 - \lvert(-24) + (-40)\rvert$ $= 77 - \lvert-64\rvert$ $= 77 - 64$ $= 13$ Note: Absolute value (the value of $\lvert x\rvert$) is the value of a number without regard to its sign.

Question No.	Answer	Detailed Explanation
6	D	$21 - |8 - (-5)(7)| - 54$ $= 21 - |8 - (-35)| - 54$ $= 21 - |8 + 35| - 54$ $= 21 - |43| - 54$ $= 21 - (43) - 54$ $= 21 + (-43) - 54$ $= -22 - 54$ $= -22 + (-54)$ $= -76$ Note: Absolute value (the value of $|x|$) is the value of a number without regard to its sign.
7	C	Two values could satisfy the equation. For k = 2, $|9 - 3*2| = |9 - 6| = 3$ For k = 4, $|9 - 3*4| = |9 - 12| = |-3| = 3$
8	A	The absolute value of a non-zero number is always positive.
9	C	The absolute value of a number is always positive. In this case, however, there is a negative sign outside the absolute value bars, which stands for -1, which is to be multiplied by $|24|$. Thus, $-1*24 = -24$
10	C	$|45 - 75|$ $= |-30|$ $= 30$

Rational Numbers in Context (6.NS.C.7.B)

1	A	-7 is greater than -12 so Xavier has the higher score.
2	A	Get a common denominator and compare fractions. The GCF between 6 and 12 is 12. Kelly: $\frac{5}{6} \times \frac{2}{2} = \frac{10}{12}$ Helen: $\frac{9}{12}$ Kelly has read more because $\frac{10}{12} > \frac{9}{12}$ or $\frac{5}{6} > \frac{9}{12}$.
3	D	$-52°F > -80°F$ $-52°F$ is 52 degrees below zero whereas $-80°F$ is 80 degrees below zero.

Question No.	Answer	Detailed Explanation
4	B	Rewrite these three numbers all as decimals or fractions. $1\frac{3}{4} = 1.75$ $\frac{20}{8} = 2.5$ $1.6 = 1.6$ Now compare: $2.5 > 1.75 > 1.6$ Now rewite them in their original form: $\frac{20}{8} > 1\frac{3}{4} > 1.6$
5	D	-143 means 143 feet below sea level. -134 feet means 134 feet below sea level. Therefore Doug dove deeper because he went further below sea level than Sissy.
6	C	Rewrite these three numbers all as decimals or fractions. $\frac{4}{15} = 0.27$ $0.4 = 0.4$ $\frac{1}{3} = 0.33$ Now compare: $0.4 > 0.33 > 0.27$ Now rewite them in their original form: $0.4 > \frac{1}{3} > \frac{4}{15}$
7	C	15 feet below sea level $= -15$ feet $\frac{33}{3} = 11$ feet below sea level $= -11$ feet 11.5 feet below sea level $= -11.5$ feet The deepest school is the one with the greater negative magnitude. Therefore: -15 is deeper than -11.5 is deeper than -11 or S1 > S3 > S2.
8	A	Get a common denominator and compare fractions. The GCF between 5 and 7 is 35. $\frac{4}{5} \times \frac{7}{7} = \frac{28}{35}$ $\frac{5}{7} \times \frac{5}{5} = \frac{25}{35}$ It snowed more on the first day because $\frac{28}{35} > \frac{25}{35}$ or $\frac{4}{5} > \frac{5}{7}$..

Question No.	Answer	Detailed Explanation
9	B	First determine the balance of Molly's account: $365 − $415 = −$50 Molly has a balance of −$50 so she owes the bank $50. Bill has a balance of −$45 so he owes the bank $45. Therefore Molly owes the bank more than Bill.
10	B	Jeremiah has eaten $2\frac{5}{6}$ boxes. Farley has eaten $1\frac{15}{9} = 2\frac{6}{9}$ Since they both have eaten 2 boxes lets determine which is greater $\frac{5}{6}$ or $\frac{6}{9}$ The GCF of 6 and 9 is 18. $$\frac{5}{6} \times \frac{3}{3} = \frac{15}{18} \; ; \; \frac{6}{9} \times \frac{2}{2} = \frac{12}{18}$$ Since $\frac{15}{18} > \frac{12}{18}$ Jeremiah has eaten more, therefore Farley has more left.

Interpreting Absolute Value (6.NS.C.7.C)

1	C	The absolute value of a number is its distance from zero on a number line. Each tick on the number line is 1 unit. So the number represented by the red dot is 3. The $|3|$ is 3 because it is three units from 0.
2	C	The absolute value of a number is its distance from zero on a number line. Each tick on the number line is 0.5 units. So the number represented by the red dot is −1. The $|−1|$ is 1 because it is 1 unit from 0.
3	A	The absolute value of a number is its distance from zero on a number line. $|−27| = 27$ and $27 > 19$. Therefore $|−27| > 19$.
4	A	The absolute value of a number is its distance from zero on a number line. $− |−6|$ or the negative of the absolute value of −6 is −6 and −4 > −6. Therefore $−4 > −|−6|$.
5	B	To make the statement $−12 < x$ true, x must be greater than −12. To make the statement $|x| < 4$, x must be greater than −4 and less than 4. The only answer choice meeting these criteria is x=−1.

Question No.	Answer	Detailed Explanation
6	D	$\lvert 30 \rvert = 30$ $\lvert 24 \rvert = 24$ $\lvert -30 \rvert = 30$ Therefore both 30 and -30 have the greater absolute value.
7	D	$\lvert 215 \rvert = \$215$ $\lvert -100 \rvert = \$100$ $\lvert -250 \rvert = \$250$ $100 < 215 < 250$ or $\lvert -100 \rvert < \lvert 215 \rvert < \lvert -250 \rvert$
8	C	True. Betsy is 6.2 meters away from the center buoy and Luis is 6.5 meters away. Betsy is closer by 0.3 meters. True. Luis is 6.5 meters from the center buoy. False. Betsy is 6.2 meters away from the center buoy and Luis is 6.5 meters away. Luis is 0.3 meters farther from the buoy. True. Betsy is 6.2 meters directly south of the center buoy which is 6.5 meters directly south of Luis therefore Betsy is $6.2 + 6.5 = 12.7$ meters from Luis.
9	D	Ophelia: $157 - 145 = 12$ pounds; $\lvert 12 \rvert = 12$ Aaron: $163 - 178 = -15$ pounds: $\lvert -15 \rvert = 15$ Nathan: $217 - 205 = 12$ pounds; $\lvert 12 \rvert = 12$ Rebecca: $128 - 136 = -8$ pounds; $\lvert -8 \rvert = 8$ Rebecca has the least absolute change of 8 pounds.
10	A	New York: $108 - (-52) = 160$ Texas: $120 - (-23) = 143$ Florida: $109 - (-2) = 111$ North Carolina: $110 - (-34) = 144$ New York has experienced the broadest range of temperature extremes.

Comparisons of Absolute Value (6.NS.C.7.D)

1	B	A negative account balance indicates a debt is owed. $-\$45$ means $\$45$ is owed. $-\$5$ means $\$5$ is owed. Therefore the account balance of $-\$45$ represents the greatest debt.
2	A	The warmer temperatures are above zero. Of these two $5°F > 2°F$. Therefore the warmest temperature is 5°F above zero.
3	A	First calculate the amount Anneliese spent: $\$58.00 + \$35.00 = \$93.00$. If her account balance is $-\$142.00$ then before she spent any money she had $-\$142.00 + \$93.00 = -\$49.00$.

Question No.	Answer	Detailed Explanation
4	B	First calculate the elevation of the campground. They climbed to an elevation of 1839 feet and then descended 527 feet to the restaurant. So the restaurant is located at 1839 − 527 = 1312 feet. The campground is 264 feet higher than the restaurant and so is located at 1312 + 264 = 1576 feet. The difference in elevation between home and the campground is 1576 − 439 = 1137 feet.
5	C	Find the total debits and credits. Debits: (2*$5.99) + $28.00 = $39.98 Credits: $46.50 Now determine the account balance: $165.00 − $39.98 + $46.50 = $171.52
6	C	Dominik: 28 feet Denzel: −5.3 feet (5.3 feet below sea level) Kaila: −7 feet (7 feet below sea level) Malik: 20 feet The lowest elevation is −7 feet where Kaila lives.
7	D	14.9 below zero = −14.9 5.1 below zero = −5.1 6 above zero = 6 15 above zero − 15 The warmest temperature is the greatest positive number or 15 degrees C in Death Valley, California.
8	D	Sato: +4 inches Ichiro: −2 inches Aran: +6 inches Mio: −2.5 inches The shortest person is Mio.
9	C	Find how much Leary spent by subtracting the change he received from the amount of money he started with. $60.00 − $3.72 = $56.28
10	B	Find the total debits and credits. Debits: $75.38 + $35.00 = $110.38 Credits: $26.16 Now determine the account balance: $629.00 − $110.38 + $26.16 =$544.78

Question No.	Answer	Detailed Explanation
		Coordinate Plane (6.NS.C.8)
1	B	Move on the x axis from −3.6, four units to the right to 0.4 (East is to the right on this map.) Then move on the y axis from −2.6, six units up to 3.4 (North is up on this map.) So, the coordinates are (0.4, 3.4).
2	D	In Quadrant II, the x value is negative and the y value is positive. The coordinates would be (−5, 8).
3	B	Absolute value is the positive value of a number. That makes the ordered pair (4, 4).
4	A	Greg starts at (0,0). He moves north (up) 5 units and west (left) 3 units. That puts him at (−3, 5).
5	C	Point C is at (−4, 5) and Point D is at (−5, −2). That means that Yolanda has to walk 1 unit west (left) and 7 units south (down).
6	D	The coordinates of Point D are (−5, −2). Absolute value is the positive value of a number. That means the absolute value is (5, 2).
7	C	The x coordinate is listed first in a coordinate pair, and the y coordinate is listed second. The absolute value of a number is its positive value. In Quadrant IV, the x coordinate is positive and the y coordinate is negative. That makes the coordinates (7, −3).
8	D	The x coordinate is listed first in a coordinate pair, and the y coordinate is listed second. The absolute value of a number is its positive value. In Quadrant II, the x coordinate is negative and the y coordinate is positive. That makes the coordinates (−3,9).
9	C	The point (0, −6) is located 6 units below the x axis. If y = 0 is sea level, the point (0, −6) is 6 units below sea level.
10	D	The x coordinate is negative and the y coordinate is positive in Quadrant II. In Quadrant IV, the x coordinate is positive, and the y coordinate is negative making the coordinates (2,−8).

Expressions & Equations

Apply and extend previous understandings of arithmetic to algebraic equations

Whole Number Exponents (6.EE.A.1)

1. **Evaluate: 5^3**

   Ⓐ 15
   Ⓑ 125
   Ⓒ 8
   Ⓓ 2

2. **Write the expression using an exponent: 2 * 2 * 2 * 2 * 2 * 2**

   Ⓐ 2 * 6
   Ⓑ 12
   Ⓒ 2^6
   Ⓓ 6^2

3. **Write the expression using an exponent: y * y * y * y**

   Ⓐ 4y
   Ⓑ y/4
   Ⓒ 4^y
   Ⓓ y^4

4. **Find the numerical value of the following expression: 11^1**

   Ⓐ 11
   Ⓑ 1
   Ⓒ 12
   Ⓓ 10

5. **Write each expression using exponents: 2 * 2 * m * m**

   Ⓐ 2(2m)
   Ⓑ 4m
   Ⓒ 2^2m^2
   Ⓓ 2/m

6. **Simplify: $4^3 * 4^2$**

    Ⓐ 20

    Ⓑ 9

    Ⓒ 4^5

    Ⓓ 20

7. **Simplify: $(b^2c)(bc^3)$**

    Ⓐ $3b/4$

    Ⓑ $3b * 4$

    Ⓒ b^3c^4

    Ⓓ bc

8. **Simplify: $(n^4x^2)^3$**

    Ⓐ $12n*6x$

    Ⓑ $n^{12}x^6$

    Ⓒ $n^{43}x^{23}$

    Ⓓ n^7x^5

9. **Simplify: $7^4/7^2$**

    Ⓐ 7^6

    Ⓑ 7^3

    Ⓒ 7^2

    Ⓓ 7^4

10. **Simplify: $[(3^5)(3^2)]^4$**

    Ⓐ 3^{28}

    Ⓑ 3^{40}

    Ⓒ 3^{10}

    Ⓓ 3^{11}

Expressions Involving Variables (6.EE.A.2)

1. When the expression 3(n + 7) is evaluated for a given value of n, the result is 33. What is the value of n?

   Ⓐ n = 4
   Ⓑ n = 5
   Ⓒ n = 21
   Ⓓ n = 120

2. Which number is acting as a coefficient in this expression? 360 + 22x − 448

   Ⓐ 360
   Ⓑ 22
   Ⓒ 448
   Ⓓ None of these

3. Evaluate the following when n = 7: 5(n − 5)

   Ⓐ 10
   Ⓑ −60
   Ⓒ 60
   Ⓓ 30

4. For which of the following values of b does the expression 4b − 9 have a value between 90 and 100?

   Ⓐ b = 104
   Ⓑ b = 26
   Ⓒ b = 48
   Ⓓ b = 24

5. Evaluate the following when n = −4: [5n − 3n] + 2n

   Ⓐ b = 16
   Ⓑ b = −20
   Ⓒ b = −16
   Ⓓ b = 0

6. **Translate the following: "Four times a number n is equal to the difference between that number and 10"**

   Ⓐ 4n = 10 − n
   Ⓑ 4 + n = 10*n
   Ⓒ 4/n = n + 10
   Ⓓ 4n = n − 10

7. **Evaluate 2y + 3y − y when y = 2.**

   Ⓐ 7
   Ⓑ 8
   Ⓒ 9
   Ⓓ 10

8. **Find the value of 2b − 4 + 6y when b = 2 and y = 3.**

   Ⓐ 16
   Ⓑ 18
   Ⓒ 0
   Ⓓ −12

9. **Translate the following, and then solve: "A number n times 16 is equal to 48."**

   Ⓐ 16n = 48, n = 4
   Ⓑ 16n = 48, n = 8
   Ⓒ 16 + n = 48, n = 32
   Ⓓ 16n = 48, n = 3

10. **For which value of x does 6x + 12 evaluate to 54?**

   Ⓐ x = 12
   Ⓑ x = 9
   Ⓒ x = 7
   Ⓓ x = 6

Identifying Expression Parts (6.EE.A.2.B)

1. **Which of the following best describes the expression 6(4−2)?**

   Ⓐ Six and the difference of four and two.
   Ⓑ The product of six and the sum of four and two.
   Ⓒ The product of six and the difference of 4 and 2.
   Ⓓ The quotient of six and the difference of four and two.

2. **Which of the following best describes the expression (8 ÷2)−10?**

   Ⓐ The quotient of eight and two subtracted from ten.
   Ⓑ Ten less than the quotient of eight and two.
   Ⓒ The difference of ten and the quotient of eight and two.
   Ⓓ The product of eight and two minus ten.

3. **What are the coefficients in the expression (2x + 15)(9x − 3)?**

   Ⓐ 2, 15, 9, −3
   Ⓑ 15, 3
   Ⓒ 15, −3
   Ⓓ 2, 9

4. **What is the value of the greatest coefficient in the expression $4a^2 + 9a - 11b^2 + 15$?**

   Ⓐ 4
   Ⓑ 9
   Ⓒ 11
   Ⓓ 15

5. **How many factors are in the following expression, 4(6 + 8) × 3(2 − 5)?**

   Ⓐ 2
   Ⓑ 4
   Ⓒ 5
   Ⓓ 6

6. Which of the following best describes the expression $4 \div (5 \times \frac{1}{2})$?

    Ⓐ The quotient of 4 and $5\frac{1}{2}$

    Ⓑ   divided by the quotient of 5 and $\frac{1}{2}$

    Ⓒ The quotient of 4 and the product of 5 and $\frac{1}{2}$

    Ⓓ The product of 5 and $\frac{1}{2}$ divided by 4.

7. Which of the following best describes the expression $(6+9) - 4$?

    Ⓐ 4 subtracted from the sum of 6 and 9.
    Ⓑ The sum of 6 and 9 subtracted from 4.
    Ⓒ The difference of 6 and 9 less 4.
    Ⓓ The sum of the quantity of 6 plus 9 and 4.

8. Which term has the smallest coefficient in the expression $8x^4 + \frac{7}{8}x^3 - 2x^2 + x$?

    Ⓐ 1st term
    Ⓑ 2nd term
    Ⓒ 3rd term
    Ⓓ 4th term

9. Which of the following accurately represents "The product of 15 and the sum of 9 and 7."?

    Ⓐ $15 + (9 \times 7)$
    Ⓑ $(15 + 9) \times 7$
    Ⓒ $15 \div (9 + 7)$
    Ⓓ $15(9 + 7)$

10. Which of the following best describes the expression $(12 \div 4) + [2 \times (-2)]$?

    Ⓐ The sum of the quotient of 12 and 4 and the product of 2 and −2.
    Ⓑ The quotient of 12 and 4 plus the difference of the product of 2 and 2.
    Ⓒ The quotient of 4 and 12 plus the product of 2 and −2.
    Ⓓ The difference of 2 times −2 and the quotient of 12 and 4.

Evaluating Expressions (6.EE.A.2.C)

1. What is the value of y in the equation $y = 3x - 13$, when x = 6?

   Ⓐ 5
   Ⓑ −4
   Ⓒ −1
   Ⓓ 4

2. What is the value of y in the equation $y = \frac{1}{4} x \div 2$, when x = 32?

   Ⓐ 2
   Ⓑ 4
   Ⓒ 8
   Ⓓ 10

3. Evaluate the following expression when a = 3 and b = −8: $3a^2 - 7b$

   Ⓐ −44
   Ⓑ −29
   Ⓒ 68
   Ⓓ 83

4. Evaluate the following expression when v = −2 and w = 155: $6v^3 + \frac{4}{5} w$

   Ⓐ 17
   Ⓑ 68
   Ⓒ 76
   Ⓓ 172

5. Evaluate the following expression when c = −3 and d = 2: $\frac{6}{d} - 10c - c^4$

   Ⓐ −108
   Ⓑ −48
   Ⓒ 39
   Ⓓ 114

6. Use the formula $V = s^3$ to find the volume of a cube with a side length of 2 cm.

   Ⓐ 4 cm²
   Ⓑ 6 cm³
   Ⓒ 8 cm³
   Ⓓ 9 cm³

7. Use the formula $A = l \times w$ to find the area of a rectangle with a length of 4.5 feet and a width of 7.3 feet.

    Ⓐ   28.15 ft²
    Ⓑ   30.35 ft²
    Ⓒ   32.35 ft²
    Ⓓ   32.85 ft²

8. Leah found a cylindrical container with a radius of $1\frac{1}{2}$ inches and a height of 11 inches. Use the formula $V = \pi r^2 h$, where $\pi = 3.14$, r is the radius and h is the height, to find the volume of the container. Round to the nearest hundredth.

    Ⓐ   51.81 in²
    Ⓑ   77.72 in²
    Ⓒ   103.62 in²
    Ⓓ   566.28 in²

9. What is the area of a square with a side of $\frac{3}{4}$ meter, $A = s^2$?

    Ⓐ $\frac{3}{16}$ m²

    Ⓑ $\frac{9}{16}$ m²

    Ⓒ $\frac{3}{4}$ m²

    Ⓓ $\frac{9}{4}$ m²

10. The area of the base of a square pyramid is 10 ft² and the height is 6 ft. What is the volume of the pyramid using the formula $V = \frac{1}{3}\beta h$, where β is the area of the base and h is the height?

    Ⓐ   20 ft³
    Ⓑ   60 ft³
    Ⓒ   200 ft³
    Ⓓ   600 ft³

Writing Equivalent Expressions (6.EE.A.3)

1. **What is an equivalent expression for 3n − 12?**

    Ⓐ 3n + 1
    Ⓑ 3n + 4
    Ⓒ 3n − 4
    Ⓓ 3(n − 4)

2. **Simplify 2n − 7n to create an equivalent expression.**

    Ⓐ 5n
    Ⓑ −5n
    Ⓒ −n(2 − 7)
    Ⓓ n(5)

3. **Which expression is equivalent to 5y + 2z − 3y + z?**

    Ⓐ z
    Ⓑ 2y + 3z
    Ⓒ yz
    Ⓓ 11yz

4. **Which expression is equivalent to 8a + 9 − 3(a + 4)?**

    Ⓐ 32 − 3a
    Ⓑ a
    Ⓒ 24a
    Ⓓ 5a − 3

5. **Which inequality has the same solution set as 3(q + 6) > 11?**

    Ⓐ 3q + 18 < 11
    Ⓑ 3q + 18 > 11
    Ⓒ 3q + 6 > 11
    Ⓓ q + 18 > 11

6. **Which inequality has the same solution set as $10 < q + q + q + q + q - 5$?**

    (A) $10 > 5q + 5$
    (B) $10 < 5q + 5$
    (C) $10 > 5q - 5$
    (D) $10 < 5q - 5$

7. **Simplify the following equation:**
    $u + u + u + u - p + p + p - r = 55$

    (A) $4u + 2p - r = 55$
    (B) $4u + p - r = 55$
    (C) $4u - 3p + r = 55$
    (D) $4u - 3p - r = 55$

8. **$36x - 12 = 108$ has the same solution(s) as** _____ .

    (A) $3(3x - 12) = 108$
    (B) $12(3x - 1) = 108$
    (C) $3(12x - 12) = 108$
    (D) $12(x - 1) = 108$

9. **Why is the expression $5(3x + 2)$ equivalent to $15x + 10$?**

    (A) The 5 has been divided into each term in parentheses.
    (B) The 5 was distributed using the Distributive Property.
    (C) The 5 was distributed using the Associative Property.
    (D) The expressions are not equal.

10. **Which expression is equivalent to $5b - 9c - 2(4b + c)$?**

    (A) $-3b + 7c$
    (B) $-3b - 7c$
    (C) $-3b - 11c$
    (D) $-3b + 11c$

Identifying Equivalent Expressions (6.EE.A.4)

1. **Which two expressions are equivalent?**

    Ⓐ (5/25)x and (1/3)x
    Ⓑ (5/25)x and (1/5)x
    Ⓒ (5/25)x and (1/4)x
    Ⓓ (5/25)x and (1/6)x

2. **Which two expressions are equivalent?**

    Ⓐ 7 + 21v and 2(5 + 3v)
    Ⓑ 7 + 21v and 3(4 + 7v)
    Ⓒ 7 + 21v and 7(1 + 3v)
    Ⓓ 7 + 21v and 7(7 +21v)

3. **Which two expressions are equivalent?**

    Ⓐ 32p/2 and 17p
    Ⓑ 32p/2 and 18p
    Ⓒ 32p/2 and 16p
    Ⓓ 32p/2 and 14p

4. **Which two expressions are equivalent?**

    Ⓐ 17(3m + 4) and 51m + 68
    Ⓑ 17(3m + 4) and 51m + 67
    Ⓒ 17(3m + 4) and 51m − 68
    Ⓓ 17(3m + 4) and 47m + 51

5. **Which two expressions are equivalent?**

    Ⓐ 64k/4 and 4k
    Ⓑ 64k/4 and 14k
    Ⓒ 64k/4 and 16k
    Ⓓ 64k/4 and 15k

6. **575d − 100 is equivalent to:**

    Ⓐ 25(23d − 4)
    Ⓑ 25(22d − 4)
    Ⓒ 25(23d + 4)
    Ⓓ 25(25d − 4)

7. **(800 + 444y)/4 is equivalent to:**

Ⓐ 200 + 44y
Ⓑ 800 + 111y
Ⓒ 200 + 111y
Ⓓ 200 − 111y

8. **5(19 −8y) is equivalent to:**

Ⓐ 95 − 35y
Ⓑ 95 + 40y
Ⓒ 85 − 40y
Ⓓ 95 − 40y

9. **The expression 3(26p − 7 + 14h) is equivalent to:**

Ⓐ 78 − 21 + 42
Ⓑ 78p + 21 + 42h
Ⓒ 78p − 21 + 42
Ⓓ 78p − 21 + 42h

10. **5(6x + 17y − 9z) is equivalent to:**

Ⓐ 30x + 82y − 45z
Ⓑ 20x + 85y − 40z
Ⓒ 30x + 85y − 45z
Ⓓ 30x − 85y + 45z

Equations and Inequalities (6.EE.B.5)

1. How many positive whole number solutions (values for x) does this inequality have?
    $x \le 20$

    Ⓐ 19
    Ⓑ 20
    Ⓒ 21
    Ⓓ Infinite

2. Which of the following correctly shows the number sentence that the following words describe? *17 is less than or equal to the product of 6 and q.*

    Ⓐ $17 \le 6q$
    Ⓑ $17 \le 6 - q$
    Ⓒ $17 < 6q$
    Ⓓ $17 \ge 6q$

3. Which of the following correctly shows the number sentence that the following words describe? *The quotient of d and 5 is 15.*

    Ⓐ $\dfrac{5}{d} = 15$

    Ⓑ $5d = 15$

    Ⓒ $\dfrac{d}{5} = 15$

    Ⓓ $d - 5 = 15$

4. Which of the following correctly shows the number sentence that the following words describe? *Three times the quantity u – 4 is less than 17*

    Ⓐ $3(u - 4) > 17$
    Ⓑ $3(u - 4) < 17$
    Ⓒ $3(u - 4) \le 17$
    Ⓓ $3(u - 4) \ge 17$

5. Which of the following correctly shows the number sentence that the following words describe? *The difference between z and the quantity 7 minus r is 54.*

    Ⓐ $z - 7 - r = 54$
    Ⓑ $z + 7 - r = 54$
    Ⓒ $z + (7 - r) = 54$
    Ⓓ $z - (7 - r) = 54$

6. Which of the following correctly shows the number sentence that the following words describe? *The square of the sum of 6 plus b is greater than 10.*

    Ⓐ $(6 + b)^2 > 10$
    Ⓑ $6^2 + b^2 > 10$
    Ⓒ $(6 + b)^2 = 10$
    Ⓓ $(6 + b)^2 < 10$

7. Which of the following correctly shows the number sentence that the following words describe? *16 less than the product of 5 and h is 21.*

    Ⓐ $16 - 5h = 21$
    Ⓑ $5h - 16 = 21$
    Ⓒ $16 - (5 + h) = 21$
    Ⓓ $16 < 5h + 21$

8. Which of the following correctly shows the number sentence that the following words describe? *8 times the quantity 2x – 7 is greater than 5 times the quantity 3x + 9.*

    Ⓐ $8(2x) - 7 > 5(3x) + 9$
    Ⓑ $8(2x - 7) \geq 5(3x + 9)$
    Ⓒ $8(2x - 7) > 5(3x + 9)$
    Ⓓ $8(2x - 7) < 5(3x + 9)$

9. A batting cage offers 8 pitches for a quarter. Raul has $1.50. Which expression could be used to calculate how many pitches Raul could get for his money?

    Ⓐ $1.50 x 8
    Ⓑ $1.50 ÷ 8
    Ⓒ ($1.50 ÷ $0.25) x 8
    Ⓓ ($1.50 ÷ $0.25)

10. For which of the following values of x is this inequality true?
    $500 - 3x > 80$

    Ⓐ x = 140
    Ⓑ x = 150
    Ⓒ x = 210
    Ⓓ x = 120

Modeling with Expressions (6.EE.B.6)

1. Roula had 117 gumballs. Amy had x less than 1/2 the amount that Roula had. Which expression shows how many gumballs Amy had?

    Ⓐ 117 − x/2
    Ⓑ (1/2)(117) − x.
    Ⓒ 117 − 2x
    Ⓓ 2x + 117

2. Benny earned $20.00 for weeding the garden. He also earned c dollars for mowing the lawn. Then he spent x dollars at the candy store. Which expression best represents this situation?

    Ⓐ $20 − c − x
    Ⓑ $20 + c − x
    Ⓒ $20 + c + x
    Ⓓ $20 + x/c

3. Clinton loves to cook. He makes a total of 23 different items. Clinton makes 6 different desserts, 12 appetizers, and x main courses. Which equation represents the total amount of food that Clinton cooked?

    Ⓐ 6 − 12 + x = 23
    Ⓑ 6 + 12 + x = 23
    Ⓒ 18 − x = 23
    Ⓓ 6 + 12 − x = 23

4. Simon has read 694 pages over the summer by reading 3 different books. He read 129 pages in the first book and he read 284 pages in the second book. Which equation shows how to figure out how many pages he read in the third book?

    Ⓐ 694 = 129 + 284 − y
    Ⓑ 694 = 413 − y
    Ⓒ 694 = 129 − 284 + y
    Ⓓ 694 = 129 + 284 + y

5. Jimmy had $45.00. He spent all of the money on a hat and a pair of jeans. He spent $19.00 on the pair of jeans and x dollars on the hat. Which of the following equations is true?

    Ⓐ $45.00 + x = $19.00
    Ⓑ $45.00 − x = $19.00
    Ⓒ x − $19.00 = $45.00
    Ⓓ $19.00 + 45.00 = x

6. **Janie had 54 stamps. She gave away t stamps. She then got back twice as many as she had given away. Which expression shows how many stamps Janie has now?**

   Ⓐ 54 − t
   Ⓑ 54 + t
   Ⓒ 54 − 2t
   Ⓓ 2t − 54

7. **The library has 2,500 books. The librarian wants to purchase x more books for the library. The director decides to buy twice as many as the librarian requested. How many books will the library have if the director purchases the number of books he wants?**

   Ⓐ 2,500 + x
   Ⓑ 2,500 + 2x
   Ⓒ 2,500 − x
   Ⓓ 2,500 − 2x

8. **There are 24 boys and 29 girls (not including Claire) attending Claire's birthday party. Which equation shows how many cupcakes Claire needs to have so that everyone, including herself, will have a cupcake?**

   Ⓐ 24 − 29 = c
   Ⓑ 24 + 29 = c
   Ⓒ 24 + 30 = c
   Ⓓ 24 + c = 29

9. **Heidi collects dolls. She had 172 dolls in her collection. Heidi acquired x more dolls from a friend. She then bought twice as many dolls from a yard sale. She now has 184 dolls in her collection. Which equation is true?**

   Ⓐ 172 + 3x = 184
   Ⓑ 172 − x + 2x = 184
   Ⓒ 172(3x) = 184
   Ⓓ 172 − 3x = 184

10. **Crystal grew 16 tomato plants. Each plant grew 10 tomatoes. She sold x of the tomatoes she had grown. Crystal has 54 tomatoes left for herself. Which equation is true?**

    Ⓐ 10(16 + x) = 54
    Ⓑ 16(10 − x) = 54
    Ⓒ 16 + x = 540
    Ⓓ 160 − x = 54

Solving One-Step Problems (6.EE.B.7)

1. **Which of the following equations describes this function?**

X	Y
13	104
17	136
20	160
9	72

Ⓐ y = 18x
Ⓑ y = x + 4
Ⓒ y = x + 32
Ⓓ y = 8x

2. **What is the value of x?**

$-7x = 56$

Ⓐ x = −7
Ⓑ x = 8
Ⓒ x = −49
Ⓓ x = −8

3. **Does this table show a linear relationship between x and y?**

X	Y
13	169
15	225
12	144
	400
16	
7	
	64

Ⓐ yes
Ⓑ no
Ⓒ yes, but only when x is positive
Ⓓ yes, but only when y is a perfect square

4. **Find the value of z:** $\dfrac{z}{5} = 20$

    Ⓐ 100
    Ⓑ 4
    Ⓒ 15
    Ⓓ 25

5. **Find the value of y:** $\dfrac{y}{3} = 12$

    Ⓐ 9
    Ⓑ 15
    Ⓒ 36
    Ⓓ 4

6. **Find the value of p: 13 + p = 39**

    Ⓐ 3
    Ⓑ 26
    Ⓒ 507
    Ⓓ 52

7. **Find the value of *w*: 6*w* = 54**

    Ⓐ 48
    Ⓑ 60
    Ⓒ 324
    Ⓓ 9

8. **Find the value of *h*: *h* − 4 = 20**

    Ⓐ 24
    Ⓑ 16
    Ⓒ −80
    Ⓓ −5

9. **Find the value of p: 72 + p = 108**

    Ⓐ p = 30
    Ⓑ p = 36
    Ⓒ p = 180
    Ⓓ p = 40

10. Find the value of n: 428 − n = 120

Ⓐ n = −548
Ⓑ n = 308
Ⓒ n = −308
Ⓓ n = 548

Representing Inequalities (6.EE.B.8)

1. A second grade class raised caterpillars. They had 12 caterpillars. Less than half of the caterpillars turned into butterflies. Which inequality shows how many caterpillars turned into butterflies?

    Ⓐ x < 6
    Ⓑ x > 6
    Ⓒ x ≤ 6
    Ⓓ x ≥ 6

2. Elliot has at least 5 favorite foods. How many favorite foods could Elliot have?

    Ⓐ 4
    Ⓑ 2
    Ⓒ none
    Ⓓ an infinite number

3. Julie has a box full of crayons. Her box of crayons has 549 crayons and at least 8 of them are red. Which inequality represents how many crayons could be red?

    Ⓐ x ≥ 549
    Ⓑ 8 ≥ x ≥ 549
    Ⓒ 8 ≥ x
    Ⓓ 8 ≤ x ≤ 549

4. Five times a number is greater than that number minus 17 is represented as
    _____ .

    Ⓐ 5x > x − 17
    Ⓑ x + 5 > x − 17
    Ⓒ 5x < x − 17
    Ⓓ 5x > x + 17

5. "A number divided by five minus five is less than negative four" is represented as
    _____ .

    Ⓐ 5x − 5 < −4
    Ⓑ x/5 − 5 < −4
    Ⓒ x/5 − 5 > −4
    Ⓓ x/5 − 5 < 4

6. **"Three times the quantity of a number times six plus three is less than 27" is represented as** _____ .

    Ⓐ 6x + 3 < 27
    Ⓑ 3(6)x + 3 < 27
    Ⓒ 3(6x + 3) < 27
    Ⓓ 6x + 3(3) < 27

7. **How would x > 3 be represented on a number line?**

    Ⓐ The number line would show an open circle over three with an arrow pointing to the left.
    Ⓑ The number line would show an open circle over three with an arrow pointing to the right.
    Ⓒ The number line would show a closed circle over three with an arrow pointing to the right.
    Ⓓ The number line would show a closed circle over three with an arrow pointing to the left.

8. **Amy and Joey each have jellybeans. The amount Amy has is 3 times the amount that Joey has. There are at least 44 jellybeans between them. Which inequality would help you figure out how many jellybeans Amy and Joey each have?**

    Ⓐ x + 3x ≥ 44
    Ⓑ 3x ≥ 44
    Ⓒ 3 + x ≥ 44
    Ⓓ x + 3x ≤ 44

9. **Sandra is a lawyer. She is working on x number of cases. She gets 8 more cases to work on. She now has more than 29 cases that she is working on. Which inequality could be used to figure out how many cases Sandra is working on?**

    Ⓐ 8x > 29
    Ⓑ x + 8 < 29
    Ⓒ x + 8 > 29
    Ⓓ x − 8 < 29

10. **There are 25 beehives on a farm. There are the same number of bees in each hive. The total number of bees on the farm is greater than 800. Which inequality could be used to figure out how many bees are in each hive?**

    Ⓐ 25/x > 800
    Ⓑ 25x < 800
    Ⓒ 25x > 800
    Ⓓ 25 + x > 800

Quantitative Relationships (6.EE.C.9)

1. Logan loves candy! He goes to the store and sees that the bulk candy is $0.79 a pound. Logan wants to buy p pounds of candy and needs to know how much money (m) he needs. Which equation would be used to figure out how much money Logan needs?

   Ⓐ m = 0.79 ÷ p
   Ⓑ m = 0.79(p)
   Ⓒ 0.79 = m(p)
   Ⓓ m = 0.79 + p

2. Logan is out of candy again! He goes to the store and sees that the bulk candy is now $0.84 a pound. Logan wants to buy 3 pounds of candy. Using the equation m = 0.84(p), figure out how much money (m) Logan needs.

   Ⓐ $1.68
   Ⓑ $2.52
   Ⓒ $2.25
   Ⓓ $2.54

3. Norman is going on a road trip. He has to purchase gas so that he can make it to his first destination. Gas is $3.55 a gallon. Norman gets g gallons. Which equation would Norman use to figure out how much money (t) it cost to get the gas?

   Ⓐ t = g(3.55)
   Ⓑ t = g ÷ 3.55
   Ⓒ t = 3.55 ÷ g
   Ⓓ t = g + 3.55

4. Norman needs to get gas again. This time, gas is $3.58 a gallon. Norman needs to get 13 gallons. Using the expression t = g(3.58), figure out how much Norman will spend (t) on gas.

   Ⓐ $46.15
   Ⓑ $39.54
   Ⓒ $46.45
   Ⓓ $46.54

5. Penny planned a picnic for her whole family. It has been very hot outside, so she needs a lot of lemonade to make sure no one is thirsty. There are 60 ounces in each bottle. Penny purchased b bottles of lemonade. She wants to figure out the total number of ounces (o) of lemonade she has. Which equation should she use?

Ⓐ b = 60(o)
Ⓑ 60 = o ÷ b
Ⓒ 60(b) = o
Ⓓ 60 = b ÷ o

6. Ethan is playing basketball in a tournament. Each game lasts 24 minutes. Ethan has 5 games to play. Which general equation could he use to help him figure out the total number of minutes that he played? Let t = the total time, g = the number of games, and m = the time per game.

Ⓐ t = g + m
Ⓑ t = g(m)
Ⓒ t = g − m
Ⓓ t = g ÷ m

7. The Spencers built a new house. They want to plant trees around their house. They want to plant 8 trees in the front yard and 17 in the backyard. The trees that the Spencer's want to plant cost $46 each. Could they use the equation t = c(n) where t is the total cost, c is the cost per tree, and n is the number of trees purchased, to figure out the cost to purchase trees for both the front and the back yards?

Ⓐ No, because the variables represent only two specific numbers that will never change.
Ⓑ Yes, because the variables represent only two specific numbers that will never change.
Ⓒ No, because the variables represent a dependent and an independent variable that can be filled in with any number.
Ⓓ Yes, because the variables represent a dependent and an independent variable that can be filled in with any number.

8. Liz is a florist. She is putting together b bouquets for a party. Each bouquet is going to have f sunflowers in it. The sunflowers cost $3 each. Which equation can Liz use to figure out the total cost (c) of the sunflowers in the bouquets?

Ⓐ c = 3(bf)
Ⓑ c = bf ÷ 3
Ⓒ c = 3b + f
Ⓓ c = 3(b + f)

9. Liz is a florist. She is putting together 5 bouquets for a party. Each bouquet is going to have 6 sunflowers in it. The sunflowers cost $3 each. Using the equation c = 3(bf), figure out how much Liz will charge for the bouquets.

Ⓐ $90
Ⓑ $30
Ⓒ $18
Ⓓ $80

10. Hen B (b) lays 4 times as many eggs as Hen A (a).

Hen A	Hen B
2	8
4	16
7	28
11	44

Which equation represents this scenario?

Ⓐ b = 4 ÷ a
Ⓑ b = 4 + a
Ⓒ b = 4a
Ⓓ b = 4a − 4

End of Expressions & Equations

Expressions & Equations

Answer Key
&
Detailed Explanations

Name: _____   Date: _____

Whole Number Exponents (6.EE.A.1)

Question No.	Answer	Detailed Explanation
1	B	The base is 5. The exponent is 3. 5 is multiplied 3 times, or 5 X 5 X 5 = 125
2	C	The base is 2. Count the factors. There are 6. 6 is the exponent. $2 * 2 * 2 * 2 * 2 * 2 = 2^6$
3	D	The base is y. Count the factors. There are 4. 4 is the exponent. $y * y * y * y = y4$
4	A	Write the factors: 11 Since 11 is the only factor, $11^1 = 11$
5	C	2^2m^2 The first base is 2. Count the number of 2s. There are 2. 2 is the exponent, so part of the expression is 2^2 The second base is m. Count the number of ms. There are 2. 2 is the exponent, so part of the expression is m^2 The full expression is written as: 2^2m^2
6	C	$4^3 * 4^2 = 4 * 4 * 4 * 4 * 4 = 4^5$
7	C	$(b^2c)(bc^3) = b^3c^4$ Keep the first base the same. b is the base. Add the exponents of b. $2 + 1 = 3$, so $b^2 * b^1 = b^3$ (Note: if no exponent is shown, it is understood to be 1.) Keep the second base the same. c is the base. Add the exponents of c: $1 + 3 = 4$ so $c^1 * c^3 = c^4$
8	B	$(n^4x^2)^3 = n^{12}x^6$ Keep the first base the same. n is the base. Multiply the inner exponent of n by the outer exponent: $4 * 3 = 12$, so n^{12} Keep the second base the same. x is the base. Multiply the inner exponent of x by the outer exponent: $2 * 3 = 6$, so x^6

Question No.	Answer	Detailed Explanation
9	C	$7^4/7^2 = 7^2$ Keep the base the same. 7 is the base. Subtract the exponent of the denominator from the exponent of the numerator. $4 - 2 = 2$. So, the expression simplifies to 7^2.
10	A	$[(3^5)(3^2)]^4 = 3^{28}$ Keep the base the same. 3 is the base. Add the exponents inside the brackets. $5 + 2 = 7$. The expression becomes $[3^7]^4$. To simplify further, multiply the exponents (since the base has a power raised to a power.) $[3^7]^4 = 3^{7\cdot4} = 3^{28}$

Expressions Involving Variables (6.EE.A.2)

1	A	Since $3(n + 7)$ is equal to 33, then $(n + 7)$ must equal 11. ($3 \times 11 = 33$) Therefore, n must equal 4, since $4 + 7 = 11$.
2	B	A number joined to a variable through multiplication is a coefficient. 22 is the coefficient of x.
3	A	When $n = 7$, the expression becomes $5(7 - 5) = 5(2) = 10$
4	B	When $b = 26$, $4b - 9 = 4(26) - 9 = 104 - 9 = 95$
5	C	When $n = -4$, the expression becomes: $= [5(-4) - 3(-4)] + 2(-4)$ $= [-20 - (-12)] - 8$ $= [-20 + 12] - 8$ $= [-8] - 8$ $= -16$
6	D	Four times a number n means to multiply the variable n by 4, 4n. is equal to means $=$ the difference between a number and 10 means to write the subtraction as is, from left to right, so $n - 10$. Therefore, $4n = n - 10$
7	B	First, combine like terms: $2y + 3y - y = 4y$ Then, substitute 2 for y: $4(2)$ $=8$

Question No.	Answer	Detailed Explanation
8	B	Substitute 2 for b and 3 for y: $2(2) - 4 + 6(3)$ $= 4 - 4 + 18$ $= 0 + 18$ $= 18$
9	D	a number n times 16 is equal to 48 a number n times 16: 16n is equal to: = $16n = 48$ $n = 3$, since $16(3) = 48$
10	C	When $x = 7$, $6x + 12 = 6(7) + 12 = 42 + 12 = 54$

Identifying Expression Parts (6.EE.A.2.B)

1	C	$(4 - 2) \rightarrow$ the difference of 4 and 2 $6(4 - 2) \rightarrow$ The product of 6 and the difference of 4 and 2.
2	B	$(8 \div 2) \rightarrow$ The quotient of 8 and 2 $- 10 \rightarrow$ less 10 or 10 subtracted from $(8 \div 2) - 10 \rightarrow$ 10 less than the quotient of 8 and 2
3	D	A coefficient is the number mutiplied by a variable. In this expression there are two variable terms, 2x and 9x. The coefficients are 2 and 9.
4	B	A coefficient is the number mutiplied by a variable. In this expression there are three variable terms, $4a^2$, $9a$ and $-11b^2$. The coefficients are 4, 9, and -11. The value of the greatest coefficient is 9.
5	B	A factor is a term being multiplied by another term. In this expression we have four factors 4, $(6+8)$, 3 and $(2-5)$
6	C	$(5 \times \frac{1}{2}) \rightarrow$ The product of 5 and $\frac{1}{2}$ $4 \div x \rightarrow$ The quotient of 4 and x $4 \div (5 \times \frac{1}{2}) \rightarrow$ The quotient of 4 and the product of 5 and $\frac{1}{2}$
7	A	$(6 + 9) \rightarrow$ The sum of 6 and 9 or 6 plus 9 $- 4 \rightarrow$ less 4 or 4 subtracted from $(6 + 9) - 4 = 4$ subtracted from the sum of 6 and 9
8	C	A coefficient is the number mutiplied by a variable. In this expression there are four variable terms, $8x^4$, $\frac{7}{8}x^3$, $-2x^2$ and x. The coefficients are 8, $\frac{7}{8}$, -2 and 1. The value of the smallest coefficient is -2. Therefore the third term has the smallest coefficient.

Date: _____

Question No.	Answer	Detailed Explanation
9	D	The sum of 9 and 7 → (9 + 7) The product of 5 and the sum of 9 and 7 → 5(9 + 7)
10	A	(12 ÷ 4) → The quotient of 12 and 4 + → sum [2 × (−2)] → the product of 2 and −2 (12 ÷ 4) + [2 × (−2)] → The sum of the quotient of 12 and 4 and the product of 2 and −2.

Evaluating Expressions (6.EE.A.2.C)

Question No.	Answer	Detailed Explanation	
1	A	$y = 3x - 13$	Original equation
		$y = 3(6) - 13$	Substitute 6 for x
		$y = 18 - 13$	Multiply
		$y = 5$	Subtract
2	B	$y = \frac{1}{4}x \div 2$	Original equation
		$y = \frac{1}{4}(32) \div 2$	Substitute 32 for x
		$y = \frac{32}{4} \div 2$	Multiply
		$y = 8 \div 2$	Simplify fraction
		$y = 4$	Divide
3	D	$3a^2 - 7b$	Original equation
		$3(3)^2 - 7(-8)$	Substitute 3 for a and −8 for b
		$3(9) - 7(-8)$	Exponents
		$27 - 7(-8)$	Multiply
		$27 - (-56)$	Multiply
		$27 + 56$	Change to adding
		83	Add
4	C	$6v^3 + \frac{4}{5}w$	Original equation
		$6(-2)^3 + \frac{4}{5}(155)$	Substitute −2 for v and 155 for w
		$6(-8) + \frac{4}{5}(155)$	Exponents
		$-48 + \frac{4}{5}(155)$	Multiply
		$-48 + 124$	Multiply
		76	Add

113

Question No.	Answer		Detailed Explanation
5	B	$\frac{6}{d} - 10c - c^4$	Original equation
		$\frac{6}{2} - 10(-3) - (-3)^4$	Substitute -3 for c and 2 for d
		$\frac{6}{2} - 10(-3) - (81)$	Exponents
		$3 - 10(-3) - (81)$	Divide
		$3 - (-30) - (81)$	Multiply
		$3 + 30 - (81)$	Change to adding
		$33 - (81)$	Add
		-48	Subtract
6	C	$V = s^3$	Formula
		$V = s^3$	Substitute 2 for s
		$V = 8 \text{ cm}^3$	Exponents
7	D	$A = l \times w$	Formula
		$A = 4.5 \times 7.3$	Substitute for l and w
		$A = 32.85 \text{ ft}^2$	Multiply
8	B	$V = \pi r^2 h$	Formula
		$V = (3.14)(1.5)^2 (11)$	Substitute values
		$V = (3.14)(2.25)(11)$	Exponents
		$V = 77.715 = 77.72 \text{ in}^2$	Multiply
9	B	$A = s^2$	Formula
		$A = (\frac{3}{4})^2$	Substitute values
		$A = \frac{9}{16} \text{ m}^2$	Exponents
10	A	$V = \frac{1}{3}\text{ßh}$	Formula
		$V = \frac{1}{3}(10)(6)$	Substitute values
		$V = 20 \text{ ft}^3$	Multiply

Writing Equivalent Expressions (6.EE.A.3)

1	D	The expression, $3(n - 4)$, is equivalent because it has the same value as the original. The GCF (Greatest Common Factor) of 3 has been factored out from each term.

Question No.	Answer	Detailed Explanation
2	B	Here, the variable is the common factor and can be factored out. $n(2 - 7)$ Then, simplify within the parentheses: $n(-5)$ Finally, use the Commutative Property to rewrite the expression, coefficient first: $-5n$
3	B	$5y + 2z - 3y + z = 2y + 3z$ Combine the like terms to simplify: $5y + 2z - 3y + z = (5y - 3y) + (2z + z) = 2y + 3z$
4	D	First use the Distributive Property to remove the parentheses: $8a + 9 - 3(a + 4) = 8a + 9 - 3a - 12$ Then, combine like terms: $8a + 9 - 3a - 12 = (8a - 3a) + (9 - 12) = 5a + (-3) = 5a - 3$
5	B	The Distributive Property states that a number outside of the parentheses should be distributed to all numbers inside the parentheses. The inequality symbol should not change. The correct answer is $3q + 18 > 11$
6	D	Combine like terms.: $q + q + q + q + q = 5q$ Replace the simplified expression into the original inequality: $10 < 5q - 5$
7	B	Combine like terms in the expression. $u + u + u + u = 4u$ $-p + p + p = p$ That makes the equation $4u + p - r = 55$
8	B	The GCF of 36 and 12 is 12. The Distributive Property states that a number outside of the parentheses should be distributed to all numbers inside the parentheses. So, $36x - 12 = 108$ can be rewritten as $12(3x - 1) = 108$
9	B	The Distributive Property states that a number outside of the parentheses should be distributed to all numbers inside the parentheses. $5(3x + 2) = (5*3x) + (5*2) = 15x + 10.$

Question No.	Answer	Detailed Explanation
10	C	First use the Distributive Property to remove the parentheses: 5b − 9c − 2(4b + c) = 5b − 9c − 8b − 2c Then, combine like terms: 5b − 9c − 8b − 2c = (5b − 8b) + (− 9c − 2c) = −3b + (−11c) = −3b − 11c

Identifying Equivalent Expressions (6.EE.A.4)

Question No.	Answer	Detailed Explanation
1	B	(5/25)x and (1/5)x are equivalent because 5/25 simplifies to 1/5. The expressions will be equivalent even if a number is substituted for x.
2	C	7 + 21v and 7(1 + 3v) are equivalent because if you distribute 7 to 1 +3v you will get 7 + 21v. The expressions will be equivalent even if a number is substituted for v.
3	C	32p/2 and 16p are equivalent because if you divide 32p by 2 you get 16p. The expressions will be equivalent even if a number is substituted for p.
4	A	17(3m + 4) and 51m + 68 are equivalent because if you distribute 17 to 3m + 4 you will get 51m + 68. The expressions will be equivalent even if a number is substituted for m.
5	C	64k/4 and 16k are equivalent because if you divide 64k by 4 you will get 16k. The expressions will be equivalent even if a number is substituted for k.
6	A	25(23d − 4) and 575d − 100 are equivalent because if you distribute 25 to 23d − 4 you will get 575d − 100. The expressions will be equivalent even if a number is substituted for d.
7	C	(800 + 444y)/4 and 200 + 111y are equivalent because if you divide 800 + 444y by 4 you will get 200 + 111y. The expressions will be equivalent even if a number is substituted y.
8	D	5(19 − 8y) and 95 − 40y are equivalent because if you distribute 5 to 19 − 8y you will get 95 − 40y. The expressions will be equivalent even if a number is substituted for y.
9	D	3(26p − 7 + 14h) and 78p − 21 + 42h are equivalent because if you distribute 3 to 26p − 7 + 14h you will get 78p − 21 + 42h. The expressions will be equivalent even if numbers are substituted for h and p.

Question No.	Answer	Detailed Explanation
10	C	5(6x + 17y − 9z) and 30x + 85y − 45z are equivalent because if you distribute 5 to 6x + 17y − 9z you will get 30x + 85y − 45z. The expressions will be equivalent even if x, y, and z are replaced with numbers.

Equations and Inequalities (6.EE.B.5)

Question No.	Answer	Detailed Explanation
1	B	x can be any whole number from 1 to 20, inclusive of 20.
2	A	"17 is less than or equal to" means $17 \leq$ "the product of 6 and q" means to multiply 6 and q, or 6q $17 \leq 6q$
3	C	"The quotient of d and 5" means to divide d by 5 "is 15" means "equals 15" $\frac{d}{5} = 15$
4	B	"Three times the quantity u − 4" means to multiply (u−4) by 3 → 3(u−4) "is less than 17" means < 17 $3(u-4) < 17$
5	D	"The difference between z and the quantity 7 minus r" means to find the difference between z and (7−r), so z − (7−r) "is 54" means equals 54 So, $z - (7 - r) = 54$
6	A	"The square of the sum of 6 plus b" means to square all of (6 + b), or $(6 + b)^2$ "is greater than 10" means "> 10" $(6 + b)^2 > 10$
7	B	"16 less than" means to "subtract 16 from some term" "the product of 5 and h" means to multiply 5 and h or "5h" "is 21" means "equals 21" $5h - 16 = 21$

Question No.	Answer	Detailed Explanation
8	C	"8 times the quantity 2x − 7" means to multiply 8 and 2x − 7, which needs to be in parentheses (as a quantity), so 8(2x − 7) "is greater than" means ">" 5 times the quantity 3x + 9" means 5 multiplied by 3x + 9, which needs to be in parentheses (as a quantity), so 5(3x+9) 8(2x − 7) > 5(3x + 9)
9	C	To find how many quarters (or the equivalent of how many quarters) Raul has, you could calculate $1.50 divided by $0.25. Then, that amount of quarters would be multiplied by 8, the number of pitches purchased with each quarter. The final expression would read: ($1.50 ÷ $0.25) x 8
10	D	Solve to find x: 500 − 3x > 80 First, subtract 500 from both sides −3x > − 420 Next, divide both sides by −3. (Don't forget to switch the inequality sign when dividing a negative in an inequality) x < 140 So, x = 120 would work as a solution. To check: 500 − 3(120) = 500 − 360 = 140, which is greater than 80.

Modeling with Expressions (6.EE.B.6)

1	B	Half of 117 can be expressed as (1/2)(117). x less than that is expressed as − x. (1/2)(117) − x.
2	B	Start with $20.00. Then add what he earned mowing the lawn (+c). Then subtract what he spent at the candy store (−x). The expression is $20 + c − x.
3	B	Clinton made a total of 23 items, so you know the expression has to equal 23. He made 6 dessert and 12 appetizers. You do not know how many main courses he made, so that is represented by x. 6 + 12 + x = 23
4	D	Simon read a total of 694 pages. Add the pages from the first book (129), the second book (284) and the third book (y) together. 694 = 129 + 284 + y

Question No.	Answer	Detailed Explanation
5	B	Subtracting the cost of the hat (x) from the total cost ($45.00) would leave the cost of the jeans ($19.00). $45.00 − x = $19.00.
6	B	Janie started with 54 stamps. Giving t stamps away means −t, so 54 − t. Then, Janie gets double t stamps back or + 2t. 54 − t + 2t. Simplified, the expression would be 54 + t.
7	B	The director wants to buy double x books, which is 2x. Add that to the 2,500 existing books to create the expression 2,500 + 2x.
8	C	The total number of cupcakes is "c." There are 24 boys and 29 girls plus Claire (to make 30). Add all of the people together to equal "c." 24 + 30 = c
9	A	Heidi had 172 dolls to start with. She acquired x more so add x to 172. She then acquired 2x more, so add that to 172 also. 172 + x + 2x = 184 Simplify the equation to be: 172 + 3x = 184
10	D	Crystal had 16 plants and each grew 10 tomatoes. To figure out how many tomatoes Crystal had originally, multiply 16 x 10 = 160. Crystal sold x tomatoes, so subtract x from 160. Crystal has 54 tomatoes left. 160 − x = 54

Solving One-Step Problems (6.EE.B.7)

1	D	Each time a number from x is multiplied by 8, the product is found in y. So, the equation is y = 8x
2	D	To find the value of x, you must isolate it. Divide each side by −7. − 7x/−7 = 56/−7 x = −8
3	B	Each value of y is the square of the corresponding x value. This is not a linear relationship

Question No.	Answer	Detailed Explanation
4	A	$\frac{z}{5} = 20$ To find the value of z, you must isolate it. Multiply each side by 5/1 to cancel out the denominator and isolate the z. z/5 * 5/1 = z 20 * 5/1 = 100 so, z = 100
5	C	$\frac{y}{3} = 12$ To find the value of y, you must isolate it. Multiply by 3 on each side to isolate the y. y/3*(3/1) = 12(3/1) y = 36
6	B	To find the value of p, you must isolate it. Subtract 13 from both sides to isolate the variable. 13 + p − 13 = 39 − 13 p = 26
7	D	To find the value of w, you must isolate it. Divide each side by the coefficient, 6, to isolate the variable. 6w/6 = 54/6 w = 9
8	A	To find the value of x, you must isolate it. Add 4 to both sides. h − 4 = 20 h − 4 + 4 = 20 + 4 h = 24
9	B	To find the value of p, you must first isolate it. Do that by subtracting 72 from both sides. 72 + p −72 = 108 −72 p = 36

Question No.	Answer	Detailed Explanation
10	B	To find the value of n, you must isolate it. Subtract 428 from both sides. $428 - n - 428 = 120 - 428$ $-n = -308$ Divide both sides by -1 $n = 308$

Representing Inequalities (6.EE.B.8)

1	A	Half of 12 is 6. Less than 6 caterpillars turned into butterflies. That means that $x < 6$
2	D	Elliot has at least 5 different favorite foods. That means that he has more than 5 favorite foods. He could have an infinite number of favorite foods because there is no constraint on the number of favorite foods he could have.
3	D	There can be no more than 549 red crayons because that is the maximum number of crayons in the box. You know there are at least 8 red crayons, which means that x is greater than or equal to 8 and less than or equal to 549. $8 \leq x \leq 549$
4	A	"Five times a number" is represented as 5x. "Is greater than" is represented as $>$. "That number minus 17" is represented as $x - 17$. $5x > x - 17$
5	B	"A number divided by 5" is represented as x/5. "Minus five" is represented as $- 5$. "Is less than" is represented as $<$. "Negative four" is represented as -4. $x/5 - 5 < -4$
6	C	"A number times six plus three" is represented as $6x + 3$. Three times that is represented as $3(6x + 3)$. "Is less than" is represented as $<$. $3(6x + 3) < 27$
7	B	An open circle means that the number the circle is over is not included in the answer. An arrow pointing to the right means "greater than". The number line would show an open circle over three with an arrow pointing to the right.

Question No.	Answer	Detailed Explanation
8	A	The number of jellybeans that Joey has is represented as x. Amy has three times that many so that is represented as 3x. Together they have at least 44. That means that their jellybeans added together are at least 44, so that is represented as x + 3x. At least means that they could have 44 or more than 44 so that is represented as > x + 3x ≥ 44
9	C	The number of cases that Sandra has is represented as x. The 8 more she gets is represented as x + 8. She has more than 29, so that is represented as > 29 x + 8 > 29
10	C	The number of bees in each hive is represented as x. There are the same number of bees in all 25 beehives. The total number of bees is represented as 25x. The total number of bees on the farm is greater than 800. Greater than is represented as > 800. 25x > 800

Quantitative Relationships (6.EE.C.9)

1	B	"m" represents the amount of money that Logan needs "p" represents the number of pounds that Logan buys The amount of money Logan needs is found by multiplying the cost of the candy by the number of pounds that Logan buys. m = 0.79(p)
2	B	m = $0.84(3) m = $2.52
3	A	"t" represents the total amount Norman spent "g" represents the number of gallons that Norman purchased To find the total amount that Norman spent, multiply the price of the gas by the total number of gallons that Norman purchased. t = g(3.55)
4	D	t = 13 gallons ($3.58) t = 13(3.58) t = 46.54 $46.54

Question No.	Answer	Detailed Explanation
5	C	"b" represents the number of bottles of lemonade "o" represents the total number of ounces of lemonade To find out the total number of ounces of lemonade that Penny purchased, multiply the number of bottles (b) by the number of ounces in each bottle, 60. 60(b) = o
6	B	"t" represents the total number of minutes Ethan played "g" represents the number of games "m" represent the number of minutes in each game To find out the total number of minutes Ethan played, multiply the number of games by the number of minutes in each game. t = g(m)
7	D	The equation t = c(n) can be used with any numbers. The equation has a dependent quantity (the number of trees) and an independent quantity (the cost of the tree). Those quantities, no matter what they are, when multiplied together will always give the total cost.
8	A	"c" represents the total cost "b" represents the number of bouquets "f" represents the number of sunflowers To find the total cost, multiply the number of bouquets by the number of sunflowers by the price of the sunflowers. c = 3(bf)
9	A	c = 3(bf) c = 3(6)(5) c = 90 $90
10	C	The number of eggs that Hen B lays depends on the number of eggs that Hen A lays. Hen B lays 4 times more than Hen A. That is represented as 4a b = 4a

Geometry

Solve real-world and mathematical problems involving area, surface area and volume

Area (6.G.A.1)

1. **What is the area of the figure below?**

    Ⓐ 16 square units
    Ⓑ 63 square units
    Ⓒ 32 square units
    Ⓓ 45 square units

2. **What is the area of the figure below?**

    Ⓐ 12 square units
    Ⓑ 24 square units
    Ⓒ 18 square units
    Ⓓ 36 square units

3. **What is the area of the figure below? (Assume that the vertical height of the parallelogram is 3 units .)**

    Ⓐ 28 square units
    Ⓑ 12 square units
    Ⓒ 14 square units
    Ⓓ 21 square units

4. The figure shows a small square inside a larger square. What is the area of the shaded portion of the figure below?

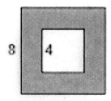

  Ⓐ 64 square units
  Ⓑ 48 square units
  Ⓒ 16 square units
  Ⓓ 80 square units

5. What is the area of the circle shown below? Round to the nearest tenth. (Let π = 3.14)

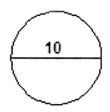

  Ⓐ 78.5 square units
  Ⓑ 314 square units
  Ⓒ 15.7 square units
  Ⓓ 31.4 square units

6. What is the area of the figure below? (Assume that the vertical height of the triangle is 2.8 units)

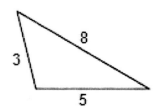

  Ⓐ 15 square units
  Ⓑ 7 square units
  Ⓒ 14 square units
  Ⓓ 40 square units

7. **What is the area of the gray portion of the figure below? The larger circle has a diameter of 24, and the smaller circle has a diameter of 12. (Let the value of π be 3.14.) Round to the nearest hundredth.**

   Ⓐ 452.16 square units
   Ⓑ 339.12 square units
   Ⓒ 113.04 square units
   Ⓓ 1,356.48 square units

8. **What is the area of the shaded part of the figure below? The radius of the larger circle is 10 and the radius of the smaller circle is 5. (Let the value of π be 3.14.) Round to the nearest tenth.**

   Ⓐ 314.0 square units
   Ⓑ 157.0 square units
   Ⓒ 31.4 square units
   Ⓓ 235.5 square units

9. **What is the area of the gray part of the squares below?.**

   Ⓐ 289 square units
   Ⓑ 81 square units
   Ⓒ 208 square units
   Ⓓ 370 square units

10. **What is the area of a triangle with a base of 20 feet and a vertical height of 40 feet?**

   Ⓐ A = 200 square ft.
   Ⓑ A = 800 square ft.
   Ⓒ A = 400 square ft.
   Ⓓ A = 600 square ft.

Surface Area and Volume (6.G.A.2)

1. **What is the total number of flat surfaces, edges, and vertices on this figure?**

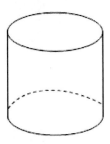

 Ⓐ zero
 Ⓑ one
 Ⓒ two
 Ⓓ three

2. **Which of the following statements is true of a rhombus?**

 Ⓐ A rhombus is a parallelogram.
 Ⓑ A rhombus is a quadrilateral.
 Ⓒ A rhombus is equilateral.
 Ⓓ All of the above are true.

3. **A cube has a volume of 1,000 cm³. What is its surface area?**

 Ⓐ 100 square cm
 Ⓑ 60 square cm
 Ⓒ 600 square cm
 Ⓓ It cannot be determined.

4. **Calculate the surface area of the cylinder shown below. Round to two decimal places.**

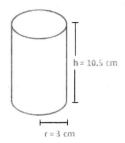

h = 10.5 cm

r = 3 cm

- Ⓐ 197.82 square cm
- Ⓑ 254.34 square cm
- Ⓒ 226.08 square cm
- Ⓓ 353.25 square cm

5. **Calculate the volume of the cylinder shown below. Round to two decimal places.**

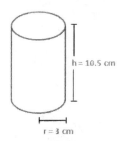

h = 10.5 cm

r = 3 cm

- Ⓐ 296.73 cubic cm
- Ⓑ 197.82 cubic cm
- Ⓒ 98.91 cubic cm
- Ⓓ 395.64 cubic cm

6. **Calculate the surface area of the box shown below.**

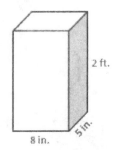

2 ft.

5 in.

8 in.

- Ⓐ 132 square inches
- Ⓑ 80 square inches
- Ⓒ 704 square inches
- Ⓓ 352 square inches

7. Find the area (to the nearest tenth of a sq. cm) of a circle whose circumference is 7π cm. (Use π = 3.14)

    Ⓐ 40.5 square cm
    Ⓑ 23.5 square cm
    Ⓒ 45 square cm
    Ⓓ 38.5 square cm

8. A single marble tile measures 25 cm by 20 cm. How many tiles will be required to cover a floor with dimensions 2 meters by 3 meters?

    Ⓐ 320 tiles
    Ⓑ 240 tiles
    Ⓒ 180 tiles
    Ⓓ 120 tiles

9. The area of one of the bases of a cylindrical can is 9π cm². The height of the can is 15 cm. What is the volume of the can? Use π = 3.14

    Ⓐ 847.8 cubic cm
    Ⓑ 135 cubic cm
    Ⓒ 423.9 cubic cm
    Ⓓ 3,815.1 cubic cm

10. What is the surface area of a rectangular box with dimensions 4 cm, 6 cm, and 10 cm?

    Ⓐ 248 square cm
    Ⓑ 240 square cm
    Ⓒ 124 square cm
    Ⓓ 224 square cm

Coordinate Geometry (6.G.A.3)

1. **The points A (0, 0), B (5, 0), C (6, 2), D (5, 5), and E (0, 5) are plotted in a coordinate grid. Describe the angles in pentagon ABCDE.**

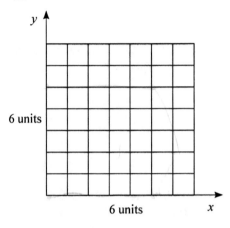

    Ⓐ 2 right angles and 3 obtuse angles
    Ⓑ 2 right angles, 2 obtuse angles, and 1 acute angle
    Ⓒ 3 right angles and 2 obtuse angles
    Ⓓ 2 right angles, 2 acute angles, and 1 obtuse angle

2. **The corners of a shape are located at (1,2), (5,2), (2,3) and (4,3) in a coordinate grid. What type of shape is it?**

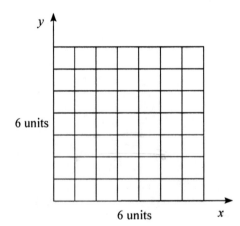

    Ⓐ square
    Ⓑ parallelogram
    Ⓒ rhombus
    Ⓓ trapezoid

3. Which of the following graphs shows a 180-degree clockwise rotation about the origin?

Ⓐ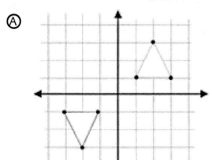

Ⓑ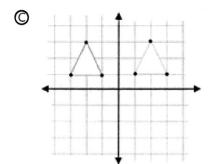

Ⓒ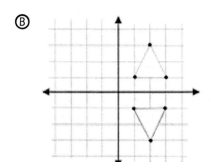

Ⓓ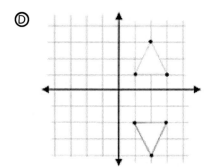

4. Identify the point on the grid below that corresponds to the ordered pair of x = 2y + 3, when y = 2.

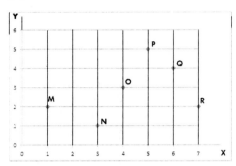

Ⓐ Point M
Ⓑ Point N
Ⓒ Point R
Ⓓ Point Q

5. What figure is formed when you draw straight line segments between the following points (in the order they are listed)? (2, –13), (2, 1), (8, 1), (8, 5), (2, 5), (2, 10), (–2, 10), (–2, 5), (–8, 5), (–8, 1), (–2, 1), (–2, –13)

Ⓐ star
Ⓑ cross
Ⓒ heart
Ⓓ boat

6. You are looking for a point on the line: y = 10 – 2x. You know that x = –1. What does y equal?

Ⓐ 10
Ⓑ 8
Ⓒ 12
Ⓓ –2

7. **Assume a function has the rule y = 2x. Which grid shows the ordered pair formed when x = 3?**

Ⓐ

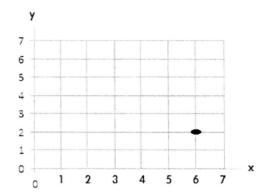

Ⓑ

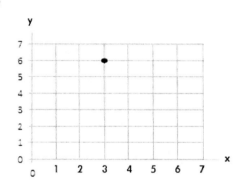

Ⓒ

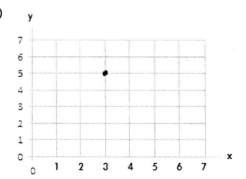

Ⓓ None of the above

8. **Which equation matches this graph?**

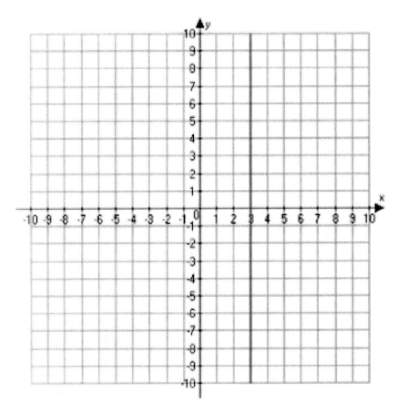

Ⓐ y = 3
Ⓑ x = 3
Ⓒ y = 0
Ⓓ x = −3

9. **What ordered pair would fit in this equation?**

y = x − 3

Ⓐ (4, 0)
Ⓑ (0, 4)
Ⓒ (4, 1)
Ⓓ (1, 4)

10. **The upper left region of the coordinate plane is Quadrant** _____

Ⓐ IV
Ⓑ I
Ⓒ II
Ⓓ III

Nets (6.G.A.4)

1. **What three dimensional figure is represented by the net below?**

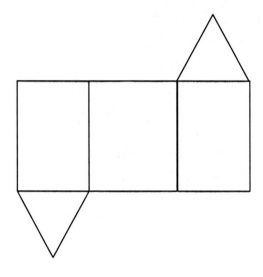

   Ⓐ Rectangular prism
   Ⓑ Rectangular pyramid
   Ⓒ Triangular prism
   Ⓓ Triangular pyramid

2. **What three dimensional figure is represented by the net below?**

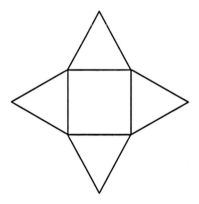

   Ⓐ Square Pyramid
   Ⓑ Cone
   Ⓒ Triangular prism
   Ⓓ Triangular pyramid

3. **What three dimensional figure is represented by the net below?**

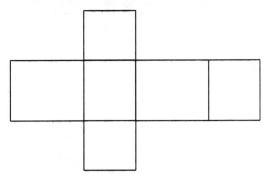

   Ⓐ **Cube**
   Ⓑ **Rectangular prism**
   Ⓒ **Cross**
   Ⓓ **Rectangular pyramid**

4. **Which of the following nets is an accurate representation of a triangular prism?**

Ⓐ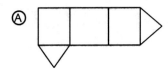

Ⓑ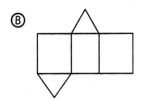

Ⓒ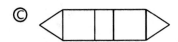

Ⓓ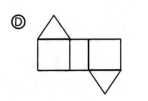

5. **What is the surface area of the figure represented below with a square center? All four triangles are the same size.**

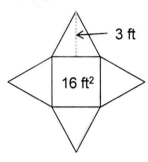

- Ⓐ 22 ft²
- Ⓑ 40 ft²
- Ⓒ 48 ft²
- Ⓓ 64 ft²

6. **What is the surface area of the figure represented below?**

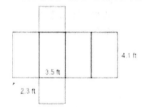

- Ⓐ 44.80 ft²
- Ⓑ 73.50 ft²
- Ⓒ 76.26 ft²
- Ⓓ 86.10 ft²

7. **A box has a width of 12 cm, a length of 30 cm and a height of 16 cm. Which expression below can be used to determine the surface area of the box?**

- Ⓐ 2(12 × 30)+2(12 × 12)+2(30 × 16)
- Ⓑ 2(12+30)+2(12+16)+2(30+16)
- Ⓒ 2(12 × 30)+2(12 × 16)+2(30 × 16)
- Ⓓ (12 × 30)+(12 × 16)+(30 × 16)

8. **Jared has a wooden box that is 9 inches in length, $6\frac{1}{2}$ inches in width, and 4.75 inches in height. What is the surface area of the wooden box?**

- Ⓐ 264.25 in²
- Ⓑ 277.875 in²
- Ⓒ 243 in²
- Ⓓ 132.125 in²

9. **What is the surface area of the cube represented in the diagram below?**

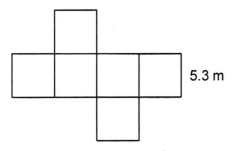

5.3 m

Ⓐ 31.8 m²
Ⓑ 63.6 m²
Ⓒ 168.54 m²
Ⓓ 339.2 m²

10. **Margot wants to paint the four walls in her room two–toned pink. The front and back walls are 10 ft by 9 ft, The two side walls are 20 ft by 9 ft. If the windows (W) take up a total of 12 sq. ft, how much surface area must Margot paint?**

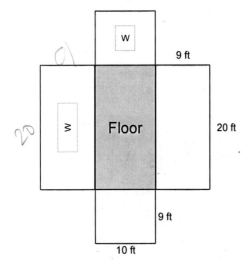

9 ft

20 ft

Floor

W

W

9 ft

10 ft

Ⓐ 528 ft²
Ⓑ 540 ft²
Ⓒ 1788 ft²
Ⓓ 1800 ft²

End of Geometry

Geometry

Answer Key
&
Detailed Explanations

Area (6.G.A.1)

Question No.	Answer	Detailed Explanation
1	B	Area of a rectangle = length x width A = 9 x 7 A = 63
2	D	Area of a square = (length of side)2 A = 6^2 A = 36
3	D	Area of a parallelogram = base x height A= 7 x 3 A = 21
4	B	First find the area of the larger square. Then find the area of the smaller square. Use the formula: A = (length of side)2 to find the area of both squares. A_{large} = 8 × 8 = 64 A_{small} = 4 × 4 = 16 Subtract the area of the smaller square from the larger square to find the area of the shaded portion (64 − 16 = 48)
5	A	Area of a circle =πr^2 A=$\pi(5^2)$ A=$\pi(25)$ A= 78.5
6	B	Area of a triangle = (1/2)bh A= (1/2) (5*2.8) A= (1/2) (14) A = 7
7	B	Area of a circle =πr^2 The area of the larger circle is A= $\pi(12^2)$ A= 452.16 Area of the smaller circle is A = $\pi(6^2)$ A= 113.04 Subtract the two areas to find the area of the shaded portion: 452.16− 113.04 = 339.12

Question No.	Answer	Detailed Explanation
8	D	Area of a circle = πr^2 For the larger circle: A= $\pi(10^2)$ = 314.0 For the smaller circle: A= $\pi(5^2)$ = 78.5 To find the area of the shaded portion, subtract the area of the smaller circle from the area of the larger circle: 314.0 − 78.5 = 235.5
9	C	Area of a square = (length of side)2 Area of the larger square is 17*17 = 289 units² Area of the smaller square is 9*9 = 81 units² To find the area of the shaded portion, subtract the area of the smaller square from the area of the larger square: 289 − 81 = 208
10	C	The formula for the area of a triangle is: A = (1/2)bh a= (1/2)(20)(40) a = (1/2)(800) a = 400 sq. ft.

Surface Area and Volume (6.G.A.2)

1	C	There are 2 flat surfaces (the top and bottom bases). An edge is a line segment where two faces meet. There are none on this figure. A vertex is the point where two edges meet. There are none on this figure.
2	D	A rhombus is a parallelogram that is a quadrilateral and is equilateral.
3	C	Since the volume of a cube is found using the formula: V = (side length)³, the length of each side of this cube would be 10 [V = (10)(10)(10)]. The surface area of a cube is found by finding the area of one face: A = (side)², then multiplying by 6, since there are 6 faces, so 6(side)². $6(10)^2$ = 6(10)(10) = 6(100) = 600

Question No.	Answer	Detailed Explanation
4	B	To find the surface area, substitute the variables with the appropriate amounts. Then, evaluate. S.A. = 2πr² + 2πrh S.A. = 2(3.14)(3)² + 2(3.14)(3)(10.5) 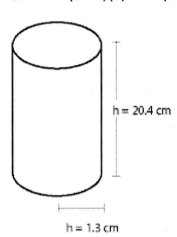 S.A. = 2(3.14)(9) + 6.28(3)(10.5) S.A. = 6.28(9) + 6.28(31.5) S.A. = 56.52 + 197.82 S.A. = 254.34
5	A	To find the volume, simply replace the variables with the appropriate values. Then, evaluate. V = πr²h V = (3.14)(3)²(10.5) V = 3.14(9)(10.5) V = 28.26(10.5) V = 296.73

Within the figure:
h = 20.4 cm

h = 1.3 cm

Question No.	Answer	Detailed Explanation
6	C	First convert 2 feet to inches. There are 12 inches per foot so 2 feet is 12 * 2 = 24 inches. To find the surface area, use the following formula: S.A. = $2(L_1W_1) + 2(L_2W_2) + 2(L_3W_3)$ where L_1W_1 represent the top and bottom of the prism L_2W_2 represent the front and back faces of the prism L_3W_3 represent the side faces of the prism. Replace the variables with the appropriate values: S.A. = 2(5*8) + 2(8*24) + 2(5*24) S.A. = 2(40) + 2(192) + 2(120) S.A. = 80 + 384 + 240 S.A. = 704 square inches
7	D	To find the area of a circle, use the formula $A = \pi r^2$ Here, however, we must first find the radius. To do so, use the following formula: $d = C/\pi$, where d = diameter and C = circumference. $d = 7\pi/\pi$ d = 7 cm Now, divide the diameter in half to find the radius. r = 7/2 = 3.5 cm A = 3.14(3.5)^2 A = 3.14(12.25) A = 38.465, and rounded to the nearest tenth, A = 38.5
8	D	First, convert meters into centimeters to standardize the measurement units. 1 meter = 100 centimeters, so 2 meters = 200 centimeters and 3 meters = 300 centimeters. To find how many tiles will be needed across, divide 300/25 = 12 To find how many tiles will be needed down, divide 200/20 = 10 Then multiply the number of tiles needed across the floor by the number of times needed down the floor. 12*10 = 120 tiles.
9	C	Use the following formula for the area of a can: $V = \pi r^2 h$ Since area = πr^2, we can substitute 9π directly into the formula V = 9π(15) V = (9)(3.14)(15) V = 423.9 cm^3

Question No.	Answer	Detailed Explanation
10	A	To find the surface area, use the following formula: S.A. = $2(L_1W_1) + 2(L_2W_2) + 2(L_3W_3)$ where L_1W_1 represent the top and bottom of the prism L_2W_2 represent the front and back faces of the prism L_3W_3 represent the side faces of the prism. Replace the variables with the appropriate values: S.A. = 2(4*6) + 2(4*10) + 2(6*10) S.A. = 2(24) + 2(40) + 2(60) S.A. = 48 + 80 + 120 S.A. = 248 sq. cm.

Coordinate Geometry (6.G.A.3)

1	A	When the points are connected, Angles A and E are right angles and Angles B, C, and D are all obtuse angles.
2	D	When the points are correctly connected, a trapezoid is created.
3	A	A 180-degree clockwise rotation about the origin will cause the top of the triangle to point to the bottom, and also the shape to shift from Quadrant I to Quadrant III.
4	C	x = 2y + 3, when y = 2 Substitute 2 for y. x = 2(2) + 3 x = 4 + 3 x = 7 This makes the ordered pair (7,2), which corresponds to point R.
5	B	When all points are correctly plotted, a cross is formed.
6	C	To determine the value of y, substitute the given value of x into the equation. y = 10 − 2x. If x = −1, then y = 10 − 2(−1) y = 10−(−2) y = 12
7	B	To determine the value of y, substitute the given value of x into the equation. y = 2x. If x = 3, then y = 2(3) y = 6 giving you the coordinate (3,6)

Question No.	Answer	Detailed Explanation
8	B	The equation, x = 3, will match the graph above, because for any given y coordinates, x will always be 3.
9	C	To determine what ordered pair would fit, substitute each x-coordinate and solve for the y-coordinate. A) (4, 0): x=4 so y = 4 − 3 = 1;The y-coordinate is 0 not 1 so this ordered pair does not fit. B) (0, 4): x=0 so y = 0 − 3 = −3;The y-coordinate is 4 not −3 so this ordered pair does not fit. C) (4,1): x=4 so y = 4 − 3 = 1; since the y-coordinate is also 1 this ordered pair fits. D) (1, 4): x=4 so y = 1 − 3 = −2;The y-coordinate is 4 not −2 so this ordered pair does not fit. Answer C is the only ordered pair that fits.
10	C	The Quadrants move counter clockwise from I to IV. The upper left region would be Quadrant II.

Nets (6.G.A.4)

1	C	Triangular prism – has three sides and two triangular bases
2	A	Square pyramid – has a square base and four triangles
3	B	Rectangular prism – has six rectangular faces (Note: this is not a cube because all the faces are not squares.)
4	D	D – Only net D is accurate as the triangular faces must be on opposite sides ajoining same size rectangular faces.

Question No.	Answer	Detailed Explanation
5	B	The surface area is the sum of the areas of all surfaces. Find the area of one triangle. Since the center is a square we know that the sides of the square are 4 ft because 4 * 4 = 16. The area of a triangle is $A = \frac{1}{2}bh$, with the base = 4 ft and the height = 3 ft, so: $$A = \frac{1}{2}bh$$ $$A = \frac{1}{2}(4)(3)$$ $A = 6$ ft² Therefore the surface area of the figure is 16 + 4(6) = 16 + 24 = 40 ft².
6	B	The surface area is the sum of the areas of all surfaces. $SA = 2(2.3 \times 3.5) + 4(3.5 \times 4.1)$ $SA = 16.1 + 57.4$ $SA = 73.5$ ft²
7	C	The surface area is the sum of the areas of all surfaces. $SA = 2(l \times w) + 2(l \times h) + 2(w \times h)$ $SA = 2(12 \times 30) + 2(12 \times 16) + 2(30 \times 16)$
8	A	The surface area is the sum of the areas of all surfaces. $SA = 2(l \times w) + 2(l \times h) + 2(w \times h)$ $SA = 2(9 \times 6.5) + 2(9 \times 4.75) +$ $\quad\quad 2(6.5 \times 4.75)$ $SA = 117 + 85.5 + 61.75$ $SA = 264.25$ in²
9	C	The surface area is the sum of the areas of all surfaces. $SA = 6s^2$ Formula $SA = 6(5.3)^2$ Substitute 5.3 for s $SA = 6(28.09)$ Exponents $SA = 168.54$ m² Multiply
10	A	The surface area is the sum of the areas of all surfaces. Front & Back Walls: A = 2(10)(9) = 180 ft² Side Walls: A = 2(20)(9) = 360 ft² Total Surface Area to Paint: 180 + 360 − 12 = 528 ft²

Statistics & Probability

Develop understanding of statistical variability

Statistical Questions (6.SP.A.1)

1. The chart below shows the participation of a sixth grade class in its school's music activites. Each student was allowed to pick one music activity.

   How many boys are in the sixth grade?

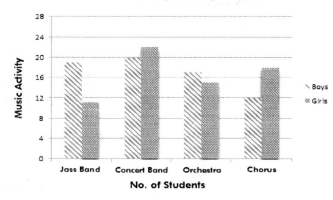

   Ⓐ 86
   Ⓑ 68
   Ⓒ 42
   Ⓓ 12

2. A student wanted to know what the sixth grade girls' favorite song was. What would be the best way to conduct a survey?

   Ⓐ Do an internet search of favorite songs of young girls.
   Ⓑ Survey the sixth grade boys.
   Ⓒ Survey the sixth grade girls.
   Ⓓ Conduct a survey at the mall.

3. **Emily wanted to know what the range of temperatures this past week was. What would be the best way of conducting her survey?**

Ⓐ Take the temperature once during the week.
Ⓑ Take the temperature only in the morning.
Ⓒ Take the temperature only in the evening.
Ⓓ Take the temperature two times a day, at the warmest and coolest times of the day.

4. **Goldie wants to find out how many presents children get for their birthdays. She surveys 11 families in the same neighborhood to find out how many presents their children get. Did Goldie get a representative sample?**

Ⓐ No because she did not ask the right questions.
Ⓑ No because she asked families in the same neighborhood who most likely have similar income levels.
Ⓒ Yes because she asked families in the same neighborhood who most likely have similar income levels.
Ⓓ Yes because she asked the right questions.

5. **Roberta is an arborist. She is studying maple trees in a specific area. Roberta wants to show a class the difference in the heights of the trees so that they can compare them. What type of graph would be best for that?**

Ⓐ Line graph
Ⓑ Picture graph
Ⓒ Circle graph
Ⓓ Bar graph

6. **Derek spends an average of 37 minutes a weekday on homework. He wants to know how much time other students in fourth grade spend on homework so he asks only students in his class. Will Derek's survey be biased?**

Ⓐ Yes because he is asking fourth graders.
Ⓑ No because he is asking fourth graders.
Ⓒ Yes because students in his class have the same amount of homework as he does.
Ⓓ No because students in his class have the same amount of homework as he does.

7. There are five cities in New York State who compete for the title of "The Snowiest City." A survey is done to find the average snowfall. The average snowfall of all 5 cities is 112.6 inches. Cities 1 and 2 both get 116 inches of snow. City 3 gets 110.4 inches of snow. City 4 gets 119.6 inches of snow. City 5 gets 101 inches of snow. Which graph would best represent this information?

    Ⓐ Picture graph
    Ⓑ Circle graph
    Ⓒ Bar graph
    Ⓓ Line graph

8. Brooke is doing a survey to find out what percentage of time families spend together doing activities. She collects all of the data about the time spent together doing those different activities and wants to put it into a graph. Which graph would be best?

    Ⓐ Circle graph
    Ⓑ Picture graph
    Ⓒ Line graph
    Ⓓ Bar graph

9. Peter and Paul are playing cards. They each randomly select 12 cards to start the game. Are the cards that they each selected a biased sample?

    Ⓐ No because they were randomly selected.
    Ⓑ Yes because they were randomly selected.
    Ⓒ Yes because they are not at least 10% of the deck.
    Ⓓ No because they are not at least 10% of the deck.

10. Cara likes to go running. She runs 4 days a week. On Monday, she runs 7 miles. On Wednesday, she runs 3.4 miles. On Friday, she runs 5 miles. On Saturday she runs 7.4 miles. Cara wants to put her data into a graph so that she can visually see the fluctuation in the miles she runs. What graph would be best?

    Ⓐ Circle graph
    Ⓑ Bar graph
    Ⓒ Line graph
    Ⓓ Picture graph

Distribution (6.SP.A.2)

1. **If the total sales for socks was $60, what is the best estimate for the total sales of pants?**

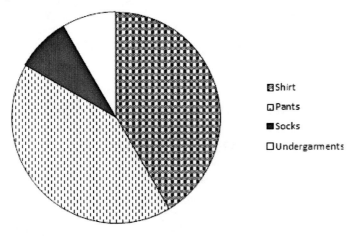

Shirt
Pants
Socks
Undergarments

Ⓐ $200
Ⓑ $300
Ⓒ $60
Ⓓ $1,000

2. **The sixth graders at Kilmer Middle School can choose to participate in one of the four music activities offered. The number of students participating in each activity is shown in the bar graph below. Use the information shown to respond to the following: How many sixth graders are in the Jazz Band?**

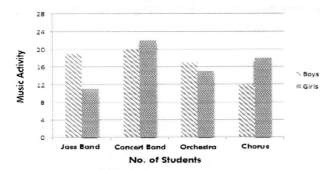

Ⓐ 19 sixth graders
Ⓑ 20 sixth graders
Ⓒ 30 sixth graders
Ⓓ 25 sixth graders

3. A .J. has downloaded 400 songs onto his computer. The songs are from a variety of genres. The circle graph below shows the breakdown (by genre) of his collection. Use the information shown to respond to the following: About how many more R + B songs than rock songs has A.J. downloaded?

A.J.'s Music Collections

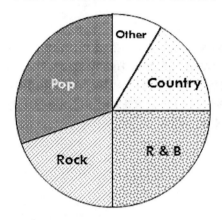

Ⓐ 20 more songs
Ⓑ 30 more songs
Ⓒ 50 more songs
Ⓓ 75 more songs

4. Colleen has to travel for work. In one week, she traveled all five work days. The shortest distance she traveled was 63 miles. The range of miles that she traveled was 98 miles. What is the longest distance that Colleen traveled for work in one week?

Ⓐ 161 miles
Ⓑ 98 miles
Ⓒ 33 miles
Ⓓ It cannot be determined.

5. Bob is a mailman. He delivers a lot of letters every day. His mail bag will only hold so many letters. The most letters that Bob has ever delivered in a day is 8,476. The range of the number of letters that Bob has ever delivered is 6,930. What is the least number of letters that Bob has ever delivered in one day?

Ⓐ 15,406 letters
Ⓑ 1,546 letters
Ⓒ 1,556 letters
Ⓓ It cannot be determined.

6. Fred goes with Katie to the art sale to try to find a gift for his mother. There are 66 pieces of art for sale. The median price is $280. Fred has $280 to spend. What is the minimum number of art pieces Fred can choose from?

Ⓐ 16
Ⓑ 33
Ⓒ 65
Ⓓ 22

7. Joann loves to watch the birds outside of her window. There is a robin that comes back every year to build a nest and lay her eggs. Joann keeps track of how many eggs the robin lays. The first year Joann keeps track, the robin lays 9 eggs and that is the least number of eggs she lays. The range of eggs that the robin has laid is one third the number of eggs from the first year. What is the most eggs the robin has ever laid?

Ⓐ 21 eggs
Ⓑ 12 eggs
Ⓒ 14 eggs
Ⓓ 27 eggs

8. Matt loves to watch movies. He has a movie collection that has 285 movies in it. The length of time of all of his movies range from 48 minutes to 3 hours and 26 minutes. What is the range of the length of time of Matt's movies?

Ⓐ 3 hours and 38 minutes
Ⓑ 1 hour and 58 minutes
Ⓒ 2 hours and 58 minutes
Ⓓ 2 hours and 38 minutes

9. Jessica is a photographer. She is putting together photo albums to show off her photographs. The first photo album has 50 pictures in it. The second photo album has 119 pictures in it. The third photo album has 174 pictures in it. The fourth photo album has 72 pictures in it. The fifth photo album has 61 pictures in it. What is the range?

Ⓐ 102
Ⓑ 174
Ⓒ 124
Ⓓ 113

10. Sam is a shrimp fisherman. In one month, Sam gets a maximum of 280 pounds of shrimp. The minimum that Sam gets is half of his maximum. What is the range in the pounds of shrimp that Sam gets?

Ⓐ 120
Ⓑ 280
Ⓒ 100
Ⓓ 140

Central Tendency (6.SP.A.3)

1   Jason was conducting a scientific experiment using bean plants. He measured the height (in centimeters) of each plant after three weeks. These were his measurements (in cm):
    12, 15, 11, 17, 19, 21, 13, 11, 16

    What is the average (mean) height? What is the median height?

    Ⓐ Mean = 15 cm, Median = 19 cm
    Ⓑ Mean = 15 cm, Median = 15 cm
    Ⓒ Mean = 19 cm, Median = 15 cm
    Ⓓ Mean = 15 cm, Median = 13 cm

2. Stacy has 60 pairs of shoes. She has shoes that have a heel height of between 1 inch and 4 inches. Stacy has 20 pairs of shoes that have a 1 inch heel, 15 pairs of shoes that have a 2 inch heel and 20 pairs of shoes that have a 3 inch heel height. What is the average heel height of all 60 pairs of Stacy's shoes?

    Ⓐ 1.3 inches
    Ⓑ 1.5 inches
    Ⓒ 2 inches
    Ⓓ 2.2 inches

3. What is the median of the following set of numbers?

    {16, −10, 13, −8, −1, 5, 7, 10}

    Ⓐ −8
    Ⓑ −1
    Ⓒ 4
    Ⓓ 6

4. Given the following set of data, is the median or the mode larger?

    {5, −10, 14, 6, 8, −2, 11, 3, 6}

    Ⓐ The mode
    Ⓑ The median
    Ⓒ They are the same
    Ⓓ You cannot figure it out

5. A = { 10, 15, 2, 14, 19, 25, 0 }

    Which of the following numbers, if added to Set A, would have the greatest effect on its median?

    Ⓐ 14
    Ⓑ 50
    Ⓒ 5
    Ⓓ 15

6. **What is the mean, median and mode for the following data?**

    {−7, 18, 29, 4, −3, 11, 22}

    Ⓐ Mean = 10
        Median = 18
        Mode = −3

    Ⓑ Mean = 11
        Median = 11
        Mode = none

    Ⓒ Mean = 10.57
        Median = 11
        Mode = none

    Ⓓ Mean = 10
        Median = 18
        Mode = −3

7. **Amanda had the following numbers: 1,2,6**

    If she added the number 3 to the list...

    Ⓐ the mean would increase
    Ⓑ the mean would decrease
    Ⓒ the median would increase
    Ⓓ the median would decrease

8. A pizza shop sells the following ice cream treats:

   Strawberry Gelato: $0.60
   Chocolate Cone: $0.80
   Lemon Ice: $0.45

   What is the average sale price for the ice cream treats?

   Ⓐ $0.62
   Ⓑ $0.45
   Ⓒ $0.60
   Ⓓ $0.80

9. Marcel has 5 stamp collections. He wants to average 35 stamps per collection. So far, he has 28, 62, 12, and 44 stamps in each collection. How many stamps does he need to have in his 5th collection to average 35 stamps?

   Ⓐ 30
   Ⓑ 29
   Ⓒ 28
   Ⓓ 27

10. How much will the mean of the following set of numbers increase by if the number 53 is added to it?

   {67, 29, 40, −12, 88, −7, 11}

   Ⓐ 53
   Ⓑ 2.768
   Ⓒ 2.571
   Ⓓ 6.625

Graphs & Charts (6.SP.B.4)

1. **Which box plot accurately represents the data set below?**

{36,24,39,42,49,33,27,54,30,45}

Ⓐ
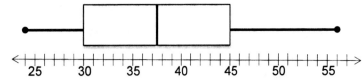

Ⓑ
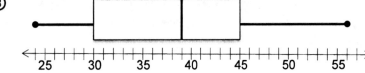

Ⓒ
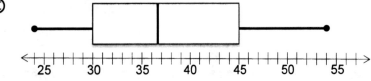

Ⓓ
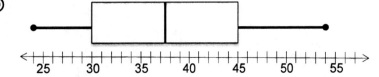

2. **Which line plot accurately represents the data below?**

{4.2,3.6,4.7,5.3,4.3,3.5,4.2,5.1,5.3,3.8,3.6 ,4.3,5.2,3.0}

Ⓐ

Ⓑ
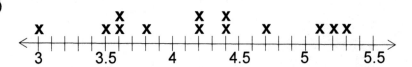

©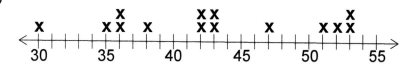

©

3. **Which histogram accurately represents the grades in Mr. Cox's class?**

{83,72,69,92,83,89,73,57,84,94,75,85,85,62,77,82,93,55,88,79,95,
85,68,77,59,84,88,98,76,73,87,79,67,84,69,76,81,99}

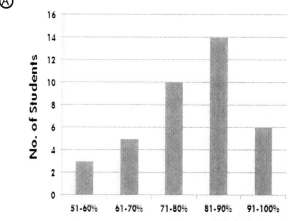

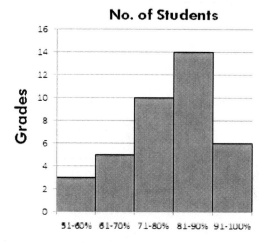

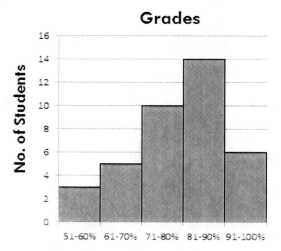

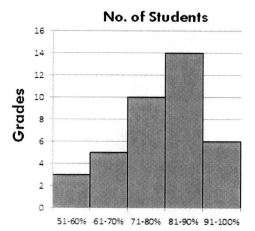

4. **Which box plot accurately represents the data set below?**

{1.2,1.2,1.3,1.5,1.5,1.5,1.6,2.6,2.8,3.0,3.0,3.1,3.3}

Ⓐ
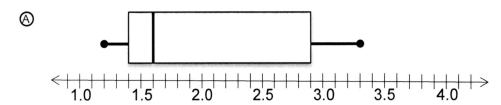

Ⓑ
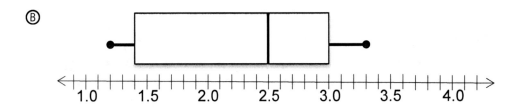

Ⓒ
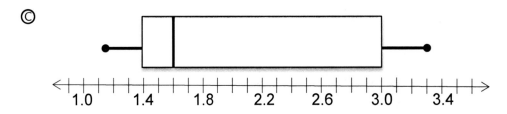

Ⓓ
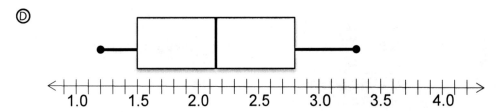

5. **Jacob belongs to a five-member track team. This week Kaleif ran 46 miles, Forest ran 38 miles, Aaron ran 56 miles, Nate ran 37 miles and Jacob ran 67 miles. Which bar graph accurately represents this data?**

Ⓐ
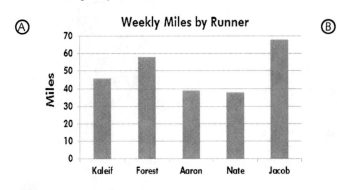

Ⓑ
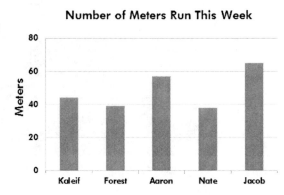

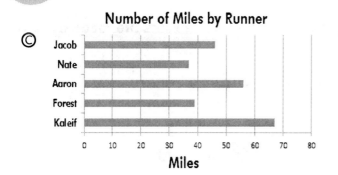

Number of Miles by Runner

Ⓒ

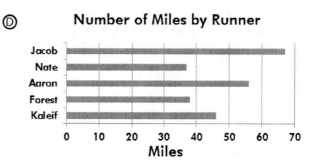

Ⓓ Number of Miles by Runner

6. **Which data set below is represented by the graph?**

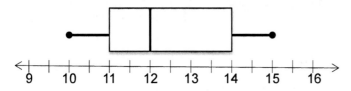

Ⓐ {10, 11, 11, 12, 12, 12, 13, 13, 14, 14, 15}
Ⓑ {10, 11, 11, 11, 12, 12, 12, 13, 13, 14, 15}
Ⓒ {10, 10, 10, 11, 11, 12, 12, 12, 13, 13, 15}
Ⓓ {10, 10, 10, 11, 11, 11, 12, 12, 12, 13, 15}

7. **Farmer John recorded the number of customers that came to his vegetable stand each day, Monday – Friday, and graphed the data below:**

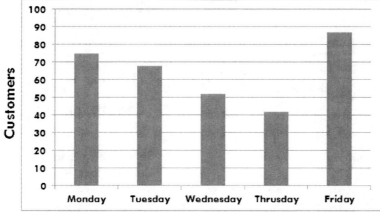

Daily Customer Count

Which data set below is represented by the graph?

Ⓐ Monday – Friday: 71, 68, 48, 42, 91
Ⓑ Monday – Friday: 75, 61, 56, 42, 87
Ⓒ Monday – Friday: 75, 68, 52, 42, 87
Ⓓ Monday – Friday: 71, 139, 191, 233, 320

8. **The local nursery sells tree saplings. The heights of the tree saplings are recorded to the nearest inch and graphed below.**

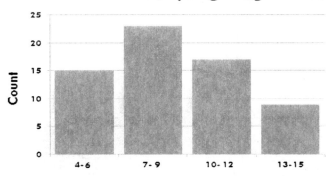

Tree Sapling Heights

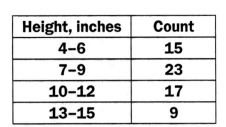

Height, inches	Count
4–6	15
7–9	23
10–12	17
13–15	9

Which one of the following statements is true.

Ⓐ The bar increments are not equal and the bars should be touching.
Ⓑ The bars should be touching and the x-axis is not properly identified.
Ⓒ The bar increments are not equal and the y-axis is not properly identified.
Ⓓ The graph is drawn properly.

9. **Which of the following graphs accurately represents the data below of string lengths to the nearest tenth of a centimeter?**

{13.2, 12.4, 13.6, 12.6, 12.2, 13.4, 13.2, 12.4, 12.8, 13.4}

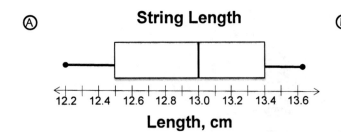

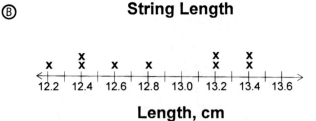

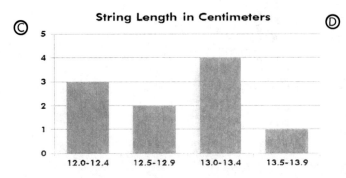

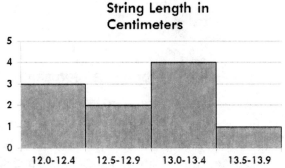

10. Which of the following statements is NOT true about graphing.

    Ⓐ On a bar graph the bars should be of equal size.

    Ⓑ On a box plot the vertical line in the Interquartile range represents the mean.

    Ⓒ In a histogram each bar represents one interval.

    Ⓓ In all graph types, the unit of measure must be included as part of the labeling.

Summarize Numerical Data Sets (6.SP.B.5)

1. **Which of the following graphs best represents the values in this table?**

X	Y
1	1
2	3
3	1
4	3

Ⓐ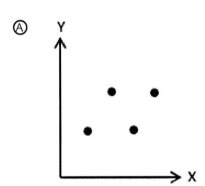

Ⓒ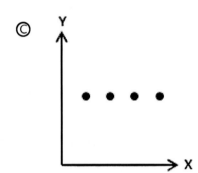

Ⓑ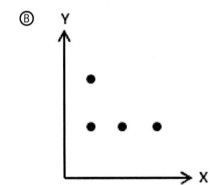

Ⓓ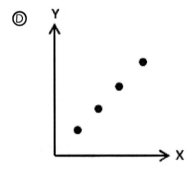

2. The results of the class' most recent science test are displayed in this histogram. Use the results to answer the question. A "passing" score is a 61 or higher.

    How many students passed the science test?

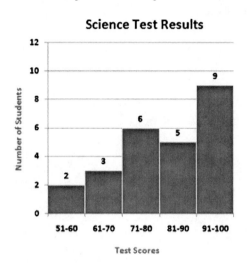

Ⓐ 3 students
Ⓑ 20 students
Ⓒ 23 students
Ⓓ 25 students

3. The results of the class' most recent science test are displayed in this histogram. Use the results to answer the question. How many students scored a 90 or below?

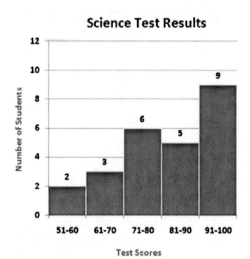

Ⓐ 5 students
Ⓑ 9 students
Ⓒ 11 students
Ⓓ 16 students

4. As part of their weather unit, the students in Mr. Green's class prepared a line graph showing the high and low temperatures recorded each day during a one-week period. Use the graph to answer the question.

On which day was the greatest range in temperature seen?

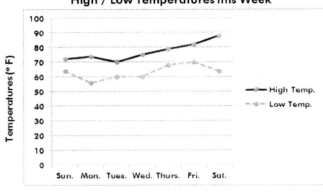

Ⓐ Sunday
Ⓑ Monday
Ⓒ Friday
Ⓓ Saturday

5. The sixth graders at Kilmer Middle School can choose to participate in one of the four music activities offered. The number of students participating in each activity is shown in the bar graph below. Use the information shown to answer the question.

There are 180 sixth graders in the school. About how many do not participate in one of the music activities?

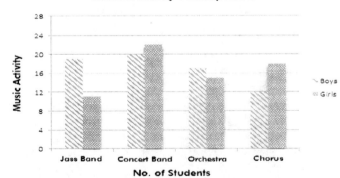

Ⓐ 66 students
Ⓑ 46 students
Ⓒ 25 students
Ⓓ 15 students

6. A.J. has downloaded 400 songs onto his computer. The songs are from a variety of genres. The circle graph below shows the breakdown of his collection by genre. Use the information shown to answer the question.

Which two genres together make up more than half of A.J.'s collection?

A.J.'s Music Collections

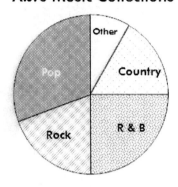

Ⓐ R + B and Country
Ⓑ Country and Rock
Ⓒ Rock and Pop
Ⓓ Pop and R + B

7. The results of the class' most recent science test are displayed in this histogram. Use the results to answer the question.

What percentage of the class scored an 81-90 on the test?

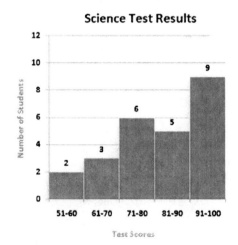

Ⓐ 5%
Ⓑ 20%
Ⓒ 25%
Ⓓ 30%

8. **How much of the graph do undergarments and socks make up together?**

Clothing Sales Breakdown

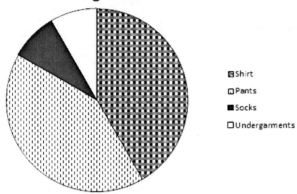

⊠ Shirt
☐ Pants
■ Socks
☐ Undergarments

Ⓐ less than 5%
Ⓑ between 5% and 10%
Ⓒ between 10% and 25%
Ⓓ more than 25%

9. **As part of their weather unit, the students in Mr. Green's class prepared a line graph showing the high and low temperatures recorded each day during a one-week period. Use the graph to answer the question.**

What percentage of the days had a temperature 80 degrees or higher? Round to the nearest tenth.

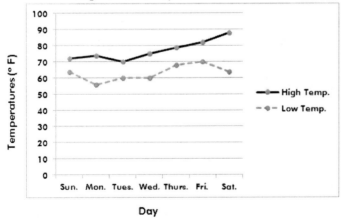

Ⓐ 14.3%
Ⓑ 42.9%
Ⓒ 28.6%
Ⓓ None of these

10. The sixth graders at Kilmer Middle School can choose to participate in one of the four music activities offered. The number of students participating in each activity is shown in the bar graph below. Use the information shown to answer the question.

There are 132 students who participate in music activities. What percentage of students who participate in music activities participate in chorus or jazz band?

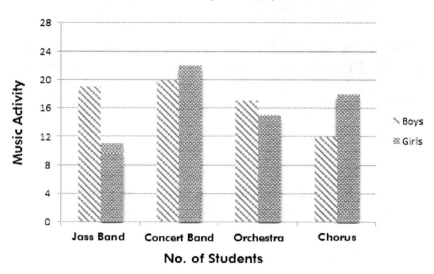

Music Activity Participation

Ⓐ 45%
Ⓑ 38%
Ⓒ 40%
Ⓓ 50%

11. A new sandwich shop just opened. It is offering a choice of ham sandwiches, chicken sandwiches, or hamburgers as the main course and French fries, potato salad, baked beans or coleslaw to go with them. How many different ways can a customer have a sandwich and one side order?

Ⓐ 6
Ⓑ 8
Ⓒ 10
Ⓓ 12

12. Karen has a set of numbers that she is working with. {6, 14, 28, 44, 2, −6} What will happen to the mean if she adds the number −8 to the set?

Ⓐ The mean will decrease.
Ⓑ The mean will increase.
Ⓒ The mean will stay the same.
Ⓓ It cannot be determined.

13. Susan goes to the store to buy supplies to make a cake. What is the average amount that Susan spent on each of the ingredients she bought?
{$4.89, $2.13, $1.10, $3.75, $0.98, $2.46}

Ⓐ $5.22
Ⓑ $2.55
Ⓒ $2.25
Ⓓ $2.50

14. Robert had an average of 87.0 on his nine math tests. His scores on the first eight tests were {92, 96, 83, 81, 94, 78, 93, 70}. What score did Robert receive on his last test?

Ⓐ 90
Ⓑ 89
Ⓒ 96
Ⓓ 87

15. Terry is playing a game with his brother and they played 5 times. The median score that Terry got was 37. The range of Terry's scores was 40. What was the lowest and highest score that Terry got?

Ⓐ 37 and 77
Ⓑ 10 and 50
Ⓒ 17 and 57
Ⓓ Not enough information is given.

16. Marcus wants to find out how many people go to the zoo on Saturdays in August. The zoo is open for 7 hours and there are 4 Saturdays in August. He counts the number of people who enter the zoo for two hours one Saturday. Will Marcus get an accurate idea of how many people go to the zoo?

Ⓐ Yes because he will get a good representative sample.
Ⓑ No because he will not get enough of a representative sample.
Ⓒ Yes because he can assume that two hours is enough time to count the number of people.
Ⓓ No because he would need to be there for less time.

17. Carl surveys his class to find out how tall his classmates are.
How many classmates did Carl survey?

Height	Number of Classmates
4'4" – 4'8"	1
4'9" – 5'	4
5'1" – 5'5"	7
5'6" – 5'11"	14
6' – 6'4"	2

Ⓐ 27
Ⓑ 28
Ⓒ 14
Ⓓ 24

18. Julie is making a quilt. She uses 7 different patterns for her quilt squares.

What percentage of Julie's quilt is not flowers or butterflies?

Squares	Number
Solid	15
Stripe	22
Flower	8
Plaid	12
Zig-Zag	13
Circles	10
Butterflies	6

Ⓐ 80%
Ⓑ 72%
Ⓒ 84%
Ⓓ 83%

19. Carl surveys his class to find out how tall his classmates are.

How many classmates are taller than 5 feet?

Ⓐ 23
Ⓑ 14
Ⓒ 27
Ⓓ 16

Height	Number of Classmates
4'4" – 4'8"	1
4'9" – 5'	4
5'1" – 5'5"	7
5'6" – 5'11"	14
6' – 6'4"	2

20. Travis is putting together outfits for work. He has 4 red shirts, 8 blue shirts and 7 white shirts. He has 6 pairs of khaki pants. What percent of Travis' possible outfits do not have red shirts?

(A) 79%
(B) 90%
(C) 78%
(D) 15%

21. Bella recorded the number of newspapers left over at the end of each day for a week. What is the average number of newspapers left each day?

3, 6, 1, 0, 2, 3, 6

(A) 2 newspapers
(B) 3 newspapers
(C) 4.3 newspapers
(D) 6 newspapers

22. Recorded below are the ages of eight friends. Which statements is true about this data set?

13, 18, 14, 12, 15, 19, 11, 15

(A) Mean > Median > Mode
(B) Mode < Mean < Median
(C) Median > Mode < Median
(D) Mode > Mean > Median

23. The line plot below represents the high temperature in °F each day of a rafting trip. What was the average high temperature during the trip?

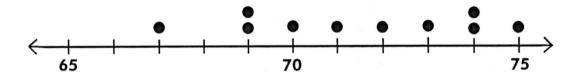

(A) 70°F
(B) 71°F
(C) 71.4°F
(D) 72°F

24. What is the mean absolute deviation of the grades Charlene received on her first ten quizzes: 83%, 92%, 76%, 87%, 89%, 96%, 88%, 91%, 79%, 99%.

   Ⓐ 4.5%
   Ⓑ 5.4%
   Ⓒ 88%
   Ⓓ 88.5%

25. The following numbers are the minutes it took Nyak to complete one lap around a certain dirt bike course. What is the mean absolute deviation of this data set?

   12.3, 12.6, 12.2, 10.9, 11.3, 10.3, 11.7, 10.7

   Ⓐ 0.7 minutes
   Ⓑ 0.75 minutes
   Ⓒ 11.1 minutes
   Ⓓ 1.5 minutes

26. The local Akita Rescue Organization has nine Akitas for adoption. Their weights are 82 lbs, 95, lbs, 130 lbs, 112 lbs, 122 lbs, 72 lbs, 86, lbs, 145 lbs, 93 lbs. What are the median (M), 1st Quartile (Q1), 3rd Quartile (Q3) and Interquartile Range (IQR) of this data set?

   Ⓐ M = 95, Q1 = 84, Q3 = 126, IQR = 42
   Ⓑ M = 95, Q1 = 85, Q3 = 126, IQR = 145
   Ⓒ M = 95, Q1 = 84, Q3 = 145, IQR = 42
   Ⓓ M = 95, Q1 = 126, Q3 = 145, IQR = 42

27. Monica is keeping track of how much money she spends on lunch each school day. This week Monica spent $5.20, $6.50, $3.75, $0.75, and $4.15. What is the median (M) and Interquartile Range (IQR) of Monica's lunch expenese?

   Ⓐ M = $4.15, IQR = $1.45
   Ⓑ M = $4.07, IQR = $1.45
   Ⓒ M = $4.07, IQR = $2.25
   Ⓓ M = $4.15, IQR = $3.60

28. Monica's friend Anna is also tracking her school lunch expenses. This week Anna spent $2.60, $0, $7.00, $4.40, $3.75. What is the difference between Monica's and Anna's average daily lunch expense?

   Monica: $5.20, $6.50, $3.75, $0.75, and $4.15

   Ⓐ $0.40
   Ⓑ $0.52
   Ⓒ $0.60
   Ⓓ $1.50

29. Zahra and Tyland are very competitive. Each have recorded their basketball scores since the beginning of the season. Which data set has a greater Interquartile Range (IQR)?

Name	Scores
Zahra	8, 10, 3, 12, 14, 11
Tyland	6, 7, 18, 12, 4, 9

  Ⓐ Zahra has the higher IQR of 4.
  Ⓑ Zahra has the higher IQR of 10.5.
  Ⓒ Tyland's IQR is 2 points higher than Zahra's.
  Ⓓ Tyland and Zahra have the same IQR of 5.

30. Jamila records the number of customers she gets each day for a week: 20, 42, 15, 23, 53, 62, 58. What is the mean absolute deviation of this data?

  Ⓐ 16.9 customers
  Ⓑ 19 customers
  Ⓒ 39 customers
  Ⓓ 47.5 customers

31. The histogram below shows the grades in Mr. Didonato's history class.

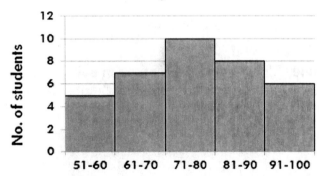

History Grades

Which of the following statements is true based on this data?

  Ⓐ The mode score is 75%.
  Ⓑ More than half the students received 81% or higher.
  Ⓒ Ten students received a 70% or lower.
  Ⓓ There is not enough information to determine the median score.

32. The bar graph below shows the different types of shoes sold by *Active Feet.*

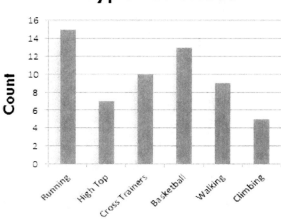

Types of Shoes

Which of the following statements is true based on this data?

Ⓐ The store carries more types of basketball and climbing shoes combined than types of running shoes.

Ⓑ The store sells more running shoes the any other shoe type.

Ⓒ The store has more types of walking shoes than cross trainers.

Ⓓ The store earns more money selling basketball shoes than walking shoes.

33. Milo and Jacque went fishing and recorded the wieght of their fish in the line plots below.

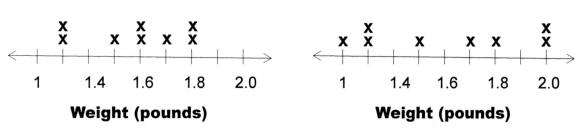

Which of the following statements is true based on this data?

Ⓐ The weights of Milo's fish have more variability.

Ⓑ The average weight of Jacque's fish is greater than the average weight of Milo's fish.

Ⓒ The weights of Jacque's fish have a greater mean absolute deviation.

Ⓓ Milo caught more fish by weight than Jacque.

34. The frequency table below records the age of the student who attended the dance.

Age	Frequency
11	17
12	20
13	9
14	8
15	12
16	4

Which of the following statements is true based on this data?

Ⓐ 80 students attended the dance.
Ⓑ The median age is 12.
Ⓒ If 3 more 16-year-olds arrive, the median age would increase.
Ⓓ More than half the students are older than the mean.

35. For twelve weeks Jeb has recorded his car's fuel mileage in miles per gallon (mpg). He has plotted these fuel mileages below in a box plot.

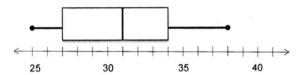

Which of the following statements is true based on this data?

Ⓐ The best fuel mileage recorded was 34 mpg.
Ⓑ Half of the data collected was between 27 mpg and 34 mpg.
Ⓒ The median fuel mileage was 30.5 mpg.
Ⓓ Three data points lie in the Interquartile range.

36. "*Forever Green*" recorded the heights of the pine trees sold this month in the box plot shown below. There are 15 data points in Quartile 3.

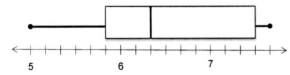

Which of the following statements is true based on this data?

Ⓐ About 25% of the trees were 7 feet 3 inches or taller.
Ⓑ There are more data points in Quartile 2 than Quartile 3.
Ⓒ The median height is 6 feet 2 inches.
Ⓓ "Forever Green" sold a total of about 60 trees this month.

37. **As part of a math project, each student in Ms. Lauzon's class recorded the number of jumping jacks they could perform in 30 seconds.**

Stem	Leaf
3	0 0 2 4 7 7
4	1 1 4 7 9
5	2 3 4 4 6 6 7
6	0 0 1 1 3 8 8
7	1 2
3|0 means 30	

Which of the following statements is true based on this data?

Ⓐ There are 21 students in Ms. Lauzon's class.
Ⓑ The median number of jumping jacks peformed was 53.
Ⓒ The range of jumping jacks is 42.
Ⓓ Seven people performed less than 40 jumping jacks.

38. **Tamryn and Clio both make beaded necklaces. The line plots below display the number of beads on ten necklaces each from Tamryn and Clio.**

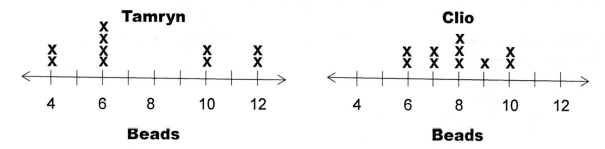

Which of the following statements is true based on this data?

Ⓐ If Tamryn added an eleventh data point of 10, the mean would not change.
Ⓑ On the average, Tamryn uses more beads than Clio.
Ⓒ Clio has more variability in the number of beads she uses.
Ⓓ If Clio added an eleventh data point of 8, the median would stay the same.

39. The histogram below shows the high temperature each day for a month.

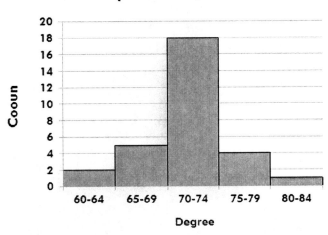

Temperature, Fahrenheit

Which of the following statements is true based on this data?

Ⓐ More than half the days had a temperature of 70–74°.
Ⓑ 20% of the days had temperatures above 74°.
Ⓒ The average temperature was 72°.
Ⓓ There were fewer days below 70° than days above 74°.

40. Marcel enjoys catching fireflies. The steam-and-leaf plot shows the number of fireflies Marcel caught on several nights.

Stem	Leaf
0	4 6 9
1	0 3 4 7 7 7 8
2	0 0 2 2 5
3	1 4
3|0 means 13	

Which of the following statements is true based on this data?

Ⓐ The median is greater than the mean.
Ⓑ The median number of fireflies caught was 17.
Ⓒ The range of fireflies is 34.
Ⓓ Marcel caught fireflies on 14 nights.

End of Statistics & Probability

Statistics & Probability

Answer Key
&
Detailed Explanations

Statistical Questions (6.SP.A.1)

Question No.	Answer	Detailed Explanation
1	B	Add the number of boys who participated in each activity to find the total number of boys. 19 + 20 + 17 + 12 = 68
2	C	When conducting a survey, it is most accurate to ask the question you are trying to answer to the focus group you are trying to reach.
3	D	Taking the temperature consistently at the warmest and coolest time of the day will provide a consistent data sample for Emily's survey.
4	B	A representative sample should be a sample of people who represent the larger population. Surveying all families in one neighborhood would not be representative because people in the same neighborhood most likely have the same income level.
5	D	A bar graph would be best because it shows the data in bars that are then very easy to compare.
6	C	Derek's survey will be biased because the students in his class have the same amount of homework as he does. To get an unbiased sample, he should ask students from all fourth grade classes.
7	C	A bar graph will show the average snowfall of each city in a bar of different colors. Those bars will be very easy to compare and will make it easy to compare the average snowfall.
8	A	A circle graph is the best graph to use when representing percentages. The circle graph would show what percentage of a family's entire day is spent doing each activity.
9	A	The randomly selected cards are not biased. A biased sample is one that contains the same or very similar items. Randomly selected samples are generally not biased.
10	C	A line graph would be best because Cara can plot the miles she runs and connect them with line segments. She will then be able to visually see the fluctuation in the miles (the trend) in the distance she runs.

Question No.	Answer	Detailed Explanation

Distribution (6.SP.A.2)

Question No.	Answer	Detailed Explanation
1	B	The Socks section has a measure of 30 degrees and the Pants section has a measure of 145 degrees. Solve for the total pant sales by setting up a proportion and solving. 145/30 = x/$60 (145)($60) = 30x 8700 = 30x 290 = x Therefore, 300 is the best estimate.
2	C	The graph shows that there are 19 boys and 11 girls participating in the Jazz Band. 19 + 11 = 30 sixth graders altogether
3	A	The R + B section is 90 degrees which is 90/360 = .25 of the total. The number of rock songs downloaded is 400 * .25 = 100 songs The Rock section is 70 degrees which is 70/360 = 0.19 which is approximately .20 of the total. The number of rock songs downloaded is 400 * .20 = 80 songs. 100 − 80 = 20. He downloaded 20 more rock songs than R & B songs.
4	A	To find the longest distance Colleen traveled, add the shortest distance to the range. 63 + 98 = 161
5	B	To find the least number of letters that Bob has ever delivered, subtract the range from the largest number. 8,476 − 6,930 = 1,546 letters
6	B	The median price is $280 and there are 66 pieces of art for sale. That means that half of the pieces of art have a price of $280 or less. Fred has $280 to spend. That means that he has at least 1/2 of the pieces of art to choose from. 66/2 = 33 There could be other art pieces priced at $280 above the median so that is why we say "at least".

Question No.	Answer	Detailed Explanation
7	B	To find the most number of eggs laid, add the least number of eggs to the range. The range is 1/3 of 9 = 3. 9 + 3 = 12 eggs
8	D	The range is the difference between the largest and smallest numbers. To find the range, subtract 48 minutes from 3 hours and 26 minutes. Change 3 hours and 26 minutes to 206 minutes 206 − 48 = 158 minutes Two hours = 120 minutes Subtract 158 − 120 to get 38 minutes remaining That makes 2 hours and 38 minutes
9	C	The range is the difference between the smallest and largest number. The largest number is 174 and the smallest number is 50. Subtract 50 from 174 to find the range. 174 − 50 = 124
10	D	To find the range, subtract the smallest number from the largest number. The largest number is 280. The smallest number is half of 280. 280/2 = 140 280 − 140 = 140

Central Tendency (6.SP.A.3)

1	B	To find the average (mean) height of the plants, the heights would first be totaled. Then the total would be divided by 9 (the number of plants in all.) 12 + 15 + 11 + 17 + 19 + 21 + 13 + 11 + 16 = 135. 135 divided by 9 equals 15. The average (mean) height is 15 centimeters. To find the median height, the numbers would be arranged in increasing order. The ordered set becomes: {11, 11, 12, 13, 15, 16, 17, 19, 21} The median is the middle value: 15 centimeters.

Question No.	Answer	Detailed Explanation
2	D	To find the average heel height, first figure out how many of each height Stacy has. Create an equation to figure out the number of 4 inch heels that Stacy has. Add 20 + 15 + 20 + x = 60 55 + x = 60 x = 5 Set up an equation to figure out the mean. You need to multiply the number of shoes and the heel height and then add them together and divide by the number of shoes. $$\frac{1(20) + 2(15) + 3(20) + 4(5) = x}{60}$$ x = 2.2
3	D	To find the median, rearrange the numbers in the data set from lowest to highest. {16, −10, 13, −8, −1, 5, 7, 10} −10, −8, −1, 5, 7, 10, 13, 16 Because this set has an even number of terms, add the two middle numbers together and divide by 2. 5 + 7 = 12 12/2 = 6 6 is the median
4	C	{5, −10, 14, 6, 8, −2, 11, 3, 6} The mode is 6 because it appears most often. To find the median, list the numbers in order from smallest to largest. −10, −2, 3, 5, 6, 6, 8, 11, 14 The median is 6 because it is in the middle.
5	C	0, 2, 10, 14, 15, 19, 25 The original median is 14. If 14 were added the median would still be 14. {0, 2, 10, 14, 14, 15, 19, 25} If 50 were added the median would be 14.5. {0, 2, 10, 14, 15, 19, 25, 50} If 5 were added the median would be 12. {0, 2, 5, 10, 14, 15, 19, 25} If 15 were added the median would be 14.5. {0, 2, 10, 14, 15, 15, 19, 25} 5 would have the greatest effect.

Question No.	Answer	Detailed Explanation
6	C	{−7, 18, 29, 4, −3, 11, 22} There is no mode because no numbers repeat To find the median, list the numbers in order from smallest to largest. −7, −3, 4, 11, 18, 22, 29 The median = 11. To find the average, add all of the numbers together and divide by 7. $\dfrac{-7 + 18 + 29 + 4 + (-3) + 11 + 22 = x}{7}$ x = 10.57
7	C	Mean for original set: 1 + 2 + 6 = 9 9/3 = 3 Median for original set: 2 Mean after adding 3: 1 + 2 + 3 + 6 = 12 12/4 = 3 Median after adding 3: 2 + 3 = 5 5/2 = 2.5
8	A	To find the average, add all of the prices together. $0.60 + $0.80 + $0.45 = $1.85 Then, divide by the number of ice cream treats. $1.85/3 = $0.616, or $0.62
9	B	For Marcel to have an average of 35 stamps, he will need to have a total of 175 stamps (5*35). Set up an equation to determine how many stamps the 5th collection should have. Let x represent the 5th collection. 28 + 62 + 12 + 44 + x = 175 146 + x = 175 Subtract 146 from both sides. 146 + x − 146 = 175 − 146 x = 29 stamps
10	B	{67, 29, 40, −12, 88, −7, 11} the mean is 30.857 {67, 29, 40, −12, 88, −7, 11, 53} the mean is 33.625 33.625 − 30.857 = 2.768

Question No.	Answer	Detailed Explanation

Graphs & Charts (6.SP.A.4)

Question No.	Answer	Detailed Explanation
1	D	To create a box plot first order the data from least to greatest. Then determine the five data points. 24 27 30 33 36 39 42 45 49 54 Minimum Point = 24 Median = $\frac{36 + 39}{2}$ = 37.5 Q1 = 30 (median of the first half of the data) Q3 = 45 (median of the second half of the data) Maximum Point = 54 Min = 24, M = 37.5, Q1 = 30, Q3 = 45, Max = 54 This matches graph D.
2	A	To create a line plot, first order the data from least to greatest. 3.0, 3.5, 3.6, 3.6, 3.8, 4.2, 4.2, 4.3, 4.3, 4.7, 5.1, 5.2, 5.3, 5.3 There should be an "x" above the number line for each data point. This matches graph A. Graph B is missing point 5.3. Graph C has a point at 5.0 and the ticks marks are not equal beyond 5.0. Graph D has the wrong scale.
3	C	To create a histogram make a frequency chart with the data. 51−60　　3 61−70　　5 71−80　　10 81−90　　14 91−100　　6 Then draw the graph with bars the height of the frequency, in equal increments with bars touching. This matches graph C. Graph A does not have touching bars. Graph B has unequal increments. Graph D is labeled incorrectly.

Question No.	Answer	Detailed Explanation
4	C	Since the data is already in order, determine the five data points. 1.2, 1.2, 1.3, 1.5, 1.5, 1.5, 1.6, 2.6, 2.8, 3.0, 3.0, 3.1, 3.3 Minimum Point = 1.2 Median = 1.6 Q1 = 1.4 (median of the first half of the data) Q3 = 3.0 (median of the second half of the data) Maximum Point = 3.3 Min = 1.2, M = 1.6, Q1 = 1.4, Q3 = 1.6, Max = 3.3 This matches graph C.
5	D	In a bar graph the height or length of the bar represents the data point. Graph D is the only correct graph. Graph A has incorrect bar heights for Forest and Aaron. Graph B is labeled with the incorrect unit of measure. Graph C has incorrect bar lengths for Jacob and Kaleif.

6	A	Determine the correct data set by calculating the five data points for each set.
		
		Set A: Min = 10, Q1 = 11, M = 12,
		    Q3 = 14, Max = 15
		Set B: Min = 10, Q1 = 11, M = 12,
		    Q3 = 13, Max = 15
		Set C: Min = 10, Q1 = 10, M = 12, Q3 = 12, Max = 15
		Set D: Min = 10, Q1 = 10, M = 11, Q3 = 12, Max = 15
		
		Graph: Min = 10, Q1 = 11, M = 12, Q3 = 14, Max = 15
		Data set A matches the graph.

7	C	Data set C matches the graph. Data points 48 and 91 do not match in data set A. Data points 61 and 56 do not match in data set B. All except the first data point in data set D do not match.

Question No.	Answer	Detailed Explanation
8	B	A: False → The bar increments are equal. B: True → The bars should be touching in a histogram and the x-axis should be labeled with the unit of measure, inches. C: False → The bar increments are equal and the y-axis is labeled accurately. D: False → The graph is not drawn properly. See B above.
9	D	First order the data points from least to greatest. 12.2, 12.4, 12.4, 12.6, 12.8, 13.2, 13.2, 13.4, 13.4, 13.6 A) Box Plot: Min = 12.2, Q1 = 12.4, M = 13, Q3 = 13.4, Max = 13.6 → Q1 is not plotted correctly. B) Line Plot: The data point 13.6 is not plotted. C) Histogram: The bars should be touching D) Hostogram: There are three numbers in the interval 12.0 – 13.4, two in the interval 12.5 – 12.9, four in the interval 13.0 – 13.4, and one in the interval 13.5 – 13.9 ⇒ This graph matches
10	B	Statement B is not true. The vertical line in the Interquartile range represents the median NOT the mean.

Summarize Numerical Data Sets (6.SP.B.5)

1	A	Using the data to create ordered pairs (x, y), the first choice is the only graph that accurately represents the ordered pairs.
2	C	To find the number of students who passed, add the number of students who scored in the ranges of 61 – 70 (3), 71 – 80 (6), 81 – 90 (5), and 91 – 100 (9). 3 + 6 + 5 + 9 = 23 students
3	D	To find the number of students who scored a 90 or below, add the number of students who scored in the ranges of 51 – 60 (2), 61 – 70 (3), 71 – 80 (6), 81 – 90 (5). 2 + 3 + 6 + 5 = 16 students
4	D	On Saturday, the greatest range was seen, with a high temperature of 88 and a low temperature of 62, making the range 26 degrees.

Question No.	Answer	Detailed Explanation
5	B	Add together all of the students and then subtract that number from the total number. Jazz: 30 students Concert Band: 42 students Orchestra: 32 students Chorus: 30 students 30 + 42 + 32 + 30 = 134 180 − 134 = 46 students do not participate in any kind of music activity
6	D	Pop and R & B together would make up more than half of the pie chart, or above 50%.
7	B	5/25 reduces to 1/5. Change 1/5 to a percentage by dividing 5 into 1. $1 \div 5 = 0.20$ Move the decimal point over two places to make a percent: 20%
8	C	Socks and undergarments together appear to take up more than a tenth, but less than a quarter, of the pie chart. The percentage would be between 10% and 25%.
9	C	Friday and Saturday both had temperatures of 80 degrees or higher. That means that 2/7 days were 80 degrees or more. To find the percentage, divide 7 into 2. $2/7 = 0.2857$ To make the decimal a percentage, move the decimal point to the right two places. $0.2857 = 28.57\%$ Round to the nearest tenth making the percentage 28.6%
10	A	Jazz Band: 19 + 11 = 30 students Chorus: 12 + 18 = 30 students 60 students participate in either jazz band or chorus. To find the percentage, divide 60 by 132. $60 \div 132 \approx 0.45$ To change the decimal to a percent, move the decimal point to the right two places. $0.45 = 45\%$
11	D	Use the counting principle to determine the number of combinations if there are 3 types of sandwiches and 4 types of sides for lunch by: $3 * 4 = 12$ There are 12 options.

Question No.	Answer	Detailed Explanation
12	A	{6, 14, 28, 44, 2, −6} the mean is 14.6 {6, 14, 28, 44, 2, −6, −8} the mean is 11.4 The mean will decrease with the addition of the number −8.
13	B	To find the average, add all of the prices together and divide by the number of ingredients, which is 6. $$\frac{\$4.89 + \$2.13 + \$1.10 + \$3.75 + \$0.98 + \$2.46}{6} = \$2.55$$
14	C	To receive an average of 87.0 on the 9 tests, Robert accumulated a total of 783 points. He scored a total of 687 on the first 8 tests. That means his score on the last test was $783 - 687 = 96$.
15	D	Just knowing the median and the range is not enough information to figure out the lowest and highest scores.
16	B	In order to get a representative sample, the person who is conducting the survey needs to get a good sample of what they are surveying. Marcus would need to spend more than 2 hours at the zoo to get a good sample. He should also go on more than one Saturday.
17	B	To figure out how many classmates Carl surveyed, add together all of the numbers in the "Number of Classmates" column. $1 + 4 + 7 + 14 + 2 = 28$
18	C	To figure out the percentage, figure out the total number of quilt squares. $15 + 22 + 8 + 12 + 13 + 10 + 6 = 86$ Figure out how many squares are not flowers or butterflies. $15 + 22 + 12 + 13 + 10 = 72$ To find the percentage, divide the number of squares that are not flowers or butterflies (72) by the total number of squares (86). $72/86 = .837$ To make the decimal into a percentage, move the decimal point to the right two places and round up. 84%
19	A	To find out how many classmates are taller than 5 feet, add together all of the classmates that are 5' 1" or taller. $7 + 14 + 2 = 23$

Question No.	Answer	Detailed Explanation
20	A	Travis has 114 possible outfits. To figure this out multiply the number of shirts by the number of khaki pants. 19 * 6 = 114 To find the outfits without a red shirt, add together the blue and white shirts (8 + 7), multiply 15 by 6 to get 90 To find the percentage, divide 90 by 114. 90/114 = .789 To make the decimal into a percentage, move the decimal point to the right two places and round up. 79%
21	B	Find the total number of newspapers leftover and divide by the number of days. $$\frac{3+6+1+0+2+3+6}{7} = \frac{21}{7} = 3 \text{ newspapers}$$
22	D	Find the mean, median, and mode of the data set and compare. Mean: $\frac{13 + 18 + 14 + 12 + 15 + 19 + 11 + 15}{8} = \frac{117}{8} = $ 14.625 Median: 11, 12, 13, 14, 15, 15, 18, 19 → 14.5 Mode: 15 Mode > Mean > Median
23	C	Add the temperatures and divide by the number of termperatures. $$\frac{67 + 69 + 69 + 70 + 71 + 72 + 73 + 74 + 74 + 75}{10} = \frac{714}{10} = $$ 71.4°F

Question No.	Answer	Detailed Explanation
24	B	First find the mean.

$$\frac{83 + 92 + 76 + 87 + 89 + 96 + 88 + 91 + 79 + 99}{10} = \frac{880}{10} = 88\%$$

Now find the distance between each number and the average.

$$|83-88|=5$$
$$|92-88|=4$$
$$|76-88|=12$$
$$|87-88|=1$$
$$|89-88|=1$$
$$|96-88|=8$$
$$|88-88|=0$$
$$|91-88|=3$$
$$|79-88|=9$$
$$|99-88|=11$$

Take the average of these differences

$$\frac{5 + 4 + 12 + 1 + 1 + 8 + 0 + 3 + 9 + 11}{10} = \frac{54}{10} = 5.4\%$$

| 25 | A | First find the mean. |

$$\frac{12.3 + 12.6 + 12.2 + 10.9 + 11.3 + 10.3 + 11.7 + 10.7}{8} = \frac{92}{8} - 11.25$$

Now find the distance between each number and the average.

$$|12.3-11.5|=0.8$$
$$|12.6-11.5|=1.1$$
$$|12.2-11.5|=0.7$$
$$|10.9-11.5|=0.6$$
$$|11.3-11.5|=0.2$$
$$|10.3-11.5|=1.2$$
$$|11.7-11.5|=0.2$$
$$|10.7-11.5|=0.8$$

Take the average of these differences

$$\frac{0.8 + 1.1 + 0.7 + 0.6 + 0.2 + 1.2 + 0.2 + 0.8}{8} = \frac{5.6}{8} = 0.7 min$$

Question No.	Answer	Detailed Explanation
26	A	Put the numbers in order from least to greatest. 72 82 86 93 95 112 122 130 145 Median = 95 $Q1 = \frac{82+86}{2} = 84$ $Q3 = \frac{122+130}{2} = 126$ IQR = 126 − 84 = 42 M = 95, Q1 = 84, Q3 = 126, IQR = 42
27	D	Put the numbers in order from least to greatest. $0.75 $3.75 $4.15 $5.20 $6.50 Median = $4.15 $Q1 = \frac{0.75+3.75}{2} = \$2.25$ $Q3 = \frac{5.20+6.50}{2} = \$5.85$ IQR = 5.85 - 2.25 = $3.60 M = $4.15, IQR = $3.60
28	B	Find the average of both and and then find the difference. Anna: $\frac{2.60+0+7.00+4.40+3.7}{5} = \frac{17.75}{5} = \$3.55$ Monica: $\frac{5.20+6.50+3.75+0.75+4.1}{5} = \frac{20.35}{5} = \$4.07$ Difference: $4.07 − $3.55 = $0.52

Question No.	Answer	Detailed Explanation
29	C	Zahra: 3 8 10 11 12 14 Median = $\frac{10+11}{2}$ = $\frac{21}{2}$ = 10.5 Q1 = 8 Q3 = 12 IQR = 12 – 8 = 4 Tyland: 4 6 7 9 12 18 Median = $\frac{7+9}{2}$ = $\frac{16}{2}$ = 8 Q1 = 6 Q3 = 12 IQR = 12 – 6 = 6 Tyland has the greater IQR of 6, which is 2 points higher tha Zahra's.
30	A	First find the mean. $\frac{20+42+15+23+53+62+58}{7}$ = $\frac{273}{7}$ = 39 Now find the distance between each number and the average. $\|20-39\|=19$ $\|42-39\|=3$ $\|15-39\|=24$ $\|23-39\|=16$ $\|53-39\|=14$ $\|62-39\|=23$ $\|58-39\|=19$ Take the average of these differences $\frac{19+3+24+16+14+23+19}{7}$ = $\frac{118}{7}$ = 16.9 *customers*
31	D	False. There is not enough information to determine the mode score. False. There are 14 out of 36 students who received an 81% or higher. This is less then a half of the students. False. Twleve students received a 70% or lower. True. We know that the median score (the average of the 18th and 19th number) falls between 71% and 80% but we do not know the specific value.

Question No.	Answer	Detailed Explanation
32	A	A) True. The store carries 13 basketball and 5 climbing shoe types for a total of 18 types. This is greater than 15 types, the number of running shoe types. B) False. The graph does not give any information about how much is sold. C) False. The store has 9 types of walking shoes and 10 types of cross trainers and thus has more types of cross trainers than walking shoes. D) False. The graph does not give any information about how much is sold.
33	C	A) False. The weights of Milo's fish have less variability has shown by the closeness of his data points. B) False. The average weight of Jack's fish is $\frac{1.3 + 1.3 + 1.5 + 1.6 + 1.6 + 1.7 + 1.8 + 1.8}{8} = \frac{12.6}{8} = 1.575$ *pounds*. The average weight of Milo's fish is $\frac{1.2 + 1.2 + 1.2 + 1.5 + 1.7 + 1.8 + 2 + 2}{8} = \frac{12.6}{8} = 1.575$ *pounds*. Hence the average weights are the same. C) True. Without any calculations you can see that Jacque's fish have weights are farther away from the mean of 1.575 lbs than Milo's fish weight data. D) False. Milo and Jacque caught the same weight in fish. See (B) above.

Question No.	Answer	Detailed Explanation
34	B	A) False. $17 + 20 + 9 + 8 + 12 + 4 = 70$ students
		B) True. $\frac{[17(11) + 20(12) + 9(13) + 8(14) + 12(15) + 4(16)]}{70} = \frac{500}{70} = 12.86 = 12$
		C) False. The current median age is the average of the 35th and 36th numbers which is $\frac{12 + 12}{2} = 12$. If three sixteen year olds arrived the median number would be the 37th number which is 12. Therefore the median age would not change.
		D) False. There are 33 students older than the mean of 12. Half of 70 would be 35 and 33 is less than 35. Therefore less than half of the students are older than the median.
35	B	A) False. The best fuel mileage recorded was 38 mpg.
		B) True. The Interquartile range is from $27 - 34$ mpg which contains 50% of the data.
		C) False. The median fuel mileage was 31 mpg.
		D) False. There are twelve data points in all which means that each quartile contains four points and the IQR contains two quartiles or eight points.
36	D	Note: Each tick mark is 2 inches.
		A) False. About 25% of the trees were 7 feet 6 inches or taller.
		B) False. Each quartile has the same number of data points.
		C) False. The median height is 6 feet 4 inches.
		D) True. If Q3 has 15 data points then so do Q1, Q2 and Q4. This makes a total of $15 * 4 = 60$ trees. We say "about" because the median may not be in a quartile.

Question No.	Answer	Detailed Explanation
37	C	A) False. There are 27 leaves which represent data for 27 students.
		B) False. The median number is the 14th number. This is 54.
		C) True. The greatest number of jumping jacks performed was 72 and the least was 30. 72 − 30 = 42.
		D) False. There are six leaves in the 3 (30) stem which represents 6 students who performed less than 40 jumping jacks.
38	D	A) False The current mean of Tamryn's data is $\frac{[2(4) + 4(6) + 2(10) + 2(12)]}{10} = \frac{76}{10} =$ 7.6. If a 10 were added the new mean would be $\frac{[2(4) + 4(6) + 2(10) + 2(12) + 10]}{11}$ $= \frac{86}{11} = 7.6$.
		B) False. Tamryn: $\frac{[2(4) + 4(6) + 2(10) + 2(12)]}{10} = \frac{76}{10} = 7.8$. Clio: $\frac{[2(6) + 2(7) + 3(8) + 9 + 2(10)]}{10} = \frac{79}{10} = 7.9$. Clio used more beads than Tamryn on the average.
		C) False. As shown by the spread of the data, Tamryn's data has more variability than Clio's.
		D) True. The median of Clio's data is 8. If Clio added another 8 the median would remain the same.
39	A	A) True. There are a total of 2 + 5 + 18 + 4 + 1 = 30 days. 18 of these days had a temperature of 70 − 74°F. This is more than half of the days.
		B) False. 4 + 1 = 5 days had temperatures above 74°F. This is $\frac{5}{30} =$ 16.6% which is less than 20%.
		C) False. There is no way to determine the average temperature without specific data points.
		D) False. There were 5 + 2 = 7 days below 74°F and 4 + 1 = 5 days above 74°F. Therefore there were more days below 74°F than days above 74°F.

Question No.	Answer	Detailed Explanation
40	B	A) False. The mean is $\dfrac{4 + 6 + 9 + 10 + 13 + 14 + 17 + 17 + 17 + 18 + 20 + 20 + 22 + 22 +}{10}$ $\dfrac{25 + 31 + 34}{10} = \dfrac{299}{17} = 17.59$. The median is 17. Therefore the mean is greater than the median. B) True. The middle number is 17. C) False. The range of fireflies is $34 - 4 = 30$. D) False. Marcel caught fireflies on 17 nights.

Notes

Notes

Lumos StepUp™ is an educational app that helps students learn and master grade-level skills in Math and English Language Arts.

The list of features includes:

- Learn Anywhere, Anytime!

- Grades 3-8 Mathematics and English Language Arts

- Get instant access to the Common Core State Standards

- One full-length sample practice test in all Grades and Subjects

- Full-length Practice Tests, Partial Tests and Standards-based Tests

- 2 Test Modes: Normal mode and Learning mode

- Learning Mode gives the user a step-by-step explanation if the answer is wrong

- Access to Online Workbooks

- Provides ability to directly scan QR Codes

- And it's completely FREE!

http://lumoslearning.com/a/stepup-app

lumoslearning

About Online Workbooks

• When you buy this book, 1 year access to online workbooks is included

• Access them anytime from a computer with an internet connection

• Adheres to the Common Core State Standards

• Includes progress reports

• Instant feedback and self-paced

• Ability to review incorrect answers

• Parents and Teachers can assist in student's learning by reviewing their areas of difficulty

Course Name: Grade 4 Math Prep

Lesson Name:	Correct	Total	% Score	Incorrect
Introduction				
Diagnostic Test		3	0%	3
Number and Numerical Operations				
Workbook - Number Sense	2	10	20%	8
Workbook - Numerical Operations	2	25	8%	23
Workbook - Estimation	1	3	33%	2
Geometry and measurement				
Workbook - Geometric Properties		6	0%	6
Workbook - Transforming Shapes				
Workbook - Coordinate Geometry	1	3	33%	2
Workbook - Units of Measurement				
Workbook - Measuring Geometric Objects	3	10	30%	7
Patterns and algebra				
Workbook - Patterns	7	10	70%	3
Workbook - Functions and relationships				

LESSON NAME: Workbook - Geometric Properties

Elapsed Time: 01 19

Question No. 2

What type of motion is being modeled here?

Select right answer
- ☐ a translation
- ☐ a rotation 90° clockwise
- ◉ a rotation 90° counter-clockwise
- ☐ a reflection

Previous question   Next question

Report Name: Missed Questions

Student Name: Lisa Colbright
Cours Name: Grade 4 Math Prep
Lesson Name: Diagnostic Test

The faces on a number cube are labeled with the numbers 1 through 6. What is the probability of rolling a number greater than 4?

Answer Explanation

(C) On a standard number cube, there are six possible outcomes. Of those outcomes, 2 of them are greater than 4. Thus, the probability of rolling a number greater than 4 is "2 out of 6" or 2/6.

A)		1/6
B)		1/3
C)	Correct Answer	2/6
D)		3/6

Lumos Learning
Developed By Expert Teachers

Grade
6

Common Core
Practice

tedBook

ENGLISH LANGUAGE ARTS

30+

⭐ Three Strands

⭐ Hundreds of Activities

SKILLS

PLUS **Online Workbooks**

Foundational Skills for
PARCC or
Smarter Balanced Tests

Available
- At Leading book stores
- Online www.LumosLearning.com

CPSIA information can be obtained
at www.ICGtesting.com
Printed in the USA
LVOW09s1645061117
555217LV00008B/755/P